Poultry Keeping
FOR BEGINNERS

DAVID KAY

DAVID & CHARLES

NEWTON ABBOT   LONDON   NORTH POMFRET(VT)   VANCOUVER

ISBN 0 7153 7395 1
Library of Congress Catalog Card Number 77–74357

Set in 12 on 13 pt Bembo
and printed in Great Britain
by Redwood Burn Limited Trowbridge
for David & Charles (Publishers) Limited
Brunel House Newton Abbot, Devon

Published in the United States of America
by David & Charles Inc
North Pomfret, Vermont 05053, USA

Published in Canada
by Douglas David & Charles Limited
1875 Welch Street North Vancouver, BC

CONTENTS

ACKNOWLEDGEMENTS

I am grateful to the many people who have helped make this book possible. My special thanks go to Thomas Alty Junior; to Wally Talbot for the photographs and Stuart Howarth for the drawings; and to Miss Jill Hargreaves for typing the manuscript. Last, but certainly not least, thanks, as always, to my wife Roberta for her patience and understanding.

PREFACE

There is usually a good reason why people want to start keeping poultry. In my own case, it goes right back to my childhood. My grandfather, a retired farmer who kept laying hens on free range, was responsible for my becoming interested in poultry when I was only three years old. He bought me three bantam hens and one cock and they were all different shapes, sizes and colours but, nevertheless, they were, from what I remember, my first proud possession. A special little cabin and run was provided for them and so too was the food. It was my responsibility to feed the bantams night and morning not to mention collecting the much sought-after eggs. Eventually, in the spring of the year a broody hen was found and some eggs were secretly incubated under her and seven chickens emerged—my flock was then eleven. As the chickens grew so did my interest.

Eventually as a young boy I started to show these bantams. First of all I showed them really as pets in classes confined to schoolchildren. Then, at the age of twelve, I made a big decision— I branched out into pure-bred stock. These birds were, of course, in my own small mind, the most valuable that ever lived and eventually birds bred from them were shown and were successful. My first effort resulted in a third prize and then gradually I crept up the ladder until my first goal was achieved—that of winning the red card—the first prize. Alongside all this I had become interested in the shapes and sizes of eggs, why people showed eggs and the points that the judges looked for and these I shall describe later in this book.

Besides helping my grandfather to look after his commercial

hens, my own showing career started to take place—a journey that was to take me through local shows to county shows and eventually to London and such shows as the International Poultry Show and Royal International Dairy Show. I have visited all these shows as an exhibitor where my birds reared under the same management and techniques described in this book have won many of the country's major prizes.

I have also had the privilege and great honour of being invited to judge these shows, and before reaching the age of 32 had judged the 'Four Royals'—Royal Show of England, Royal Highland of Scotland, Royal Welsh and Royal Ulster. In 1972, yet another goal was reached when I was invited to judge at the Canadian National Exhibition in Toronto.

In the following chapters I have endeavoured to pass on some of the knowledge which I have gained from practical experience over the years. I have tried to positively identify all the answers to the various questions that someone keeping poultry for the very first time wants to know in the hope that not only will it bring pleasure to potential poultry keepers but make their tasks easier and more successful.

I

WHY KEEP POULTRY?

There are several good reasons why many different types of people are attracted to poultry keeping. This chapter tries to identify some of the more important ones to enable you to decide whether keeping poultry is a suitable hobby for your particular demands and circumstances. Later on I shall describe the most suitable types of poultry for the beginner, and explain how best to look after them for profit and pleasure.

There are some half-dozen reasons for keeping poultry which immediately spring to mind.

The Housewife
Housewives everywhere will always support the idea of keeping poultry. All of them take pride in cracking their own home-produced eggs into a frying pan and seeing the deep-coloured yokes sizzle. Just as most people say that they prefer home-grown vegetables to deep-frozen ones, a similar parallel is often drawn between home-produced eggs and those which have been bought from the shops. Families keeping hens at home and using scraps often find that the fresh eggs have a far better colour of yolk than the bought ones. The reason for this is that these hens are being fed on natural foods rather than the artificial foods fed to birds in intensive units. However, fresh eggs are not the only reason for keeping a few hens at home. They can be kept alongside or at the rear of many homes purely for the family's own interest and for their sheer enjoyment. In addition they eat up those scraps and leftovers that would otherwise find their way to the dustbin. A strong connection is established between the housewife and her

family's hens as she sees the fresh eggs collected daily. Although it cannot be established that home-produced eggs are nutritionally better there is, nevertheless, an image of quality surrounding them.

Table Birds

Whilst some families use their scraps to supplement the feeding of laying hens others find it more useful to keep poultry for table purposes and quite naturally a different breed of fowl should be chosen. However, it must be said that there are very few people who keep poultry especially for the table. The very thoughts of killing 'William and Marmaduke' (birds tend to be christened by children as they grow) quite simply puts everyone off. In fact families have been known to stop eating poultry entirely, being put off at the very thought of having 'Marmaduke' on the table for Sunday lunch! If it is the intention to keep table poultry it must be clearly understood before the task is undertaken that the day will come when the birds have to be killed and eaten, and unless this fact can be faced then there is simply no purpose in keeping them. The management and indeed the feeding of such birds is vastly different to that of feeding ordinary laying hens and will be dealt with in a later chapter.

Young Children

Perhaps one of the most interesting reasons for keeping poultry is to provide a hobby or interest for young children. Here, the obvious type of poultry to keep are bantams. Without doubt, the keeping of bantams by young children is educational, teaching them to appreciate livestock and to understand the problems and the various habits of other animals and birds. This surely is one of the more important reasons why poultry should be kept.

An Interesting Hobby

Living in a world with a profound social atmosphere people these days seem to need to have a hobby. With more holidays and free weekends people are finding that they have more time to themselves and so have taken up hobbies which, a few years ago, they would never have even dreamt of. One of the most regular cures prescribed by the medical profession today for people who have had minor heart attacks etc, is to 'get a hobby' and the advice from many companies to employees about to retire is 'keep yourself occupied'. Indeed there are some establishments now that do provide help in advising the elderly as to ways and means of keeping themselves occupied. The keeping of poultry is naturally restricted to the amount of ground available and more thought has to be given to it than perhaps a hobby such as photography or fishing. There is, however, one distinct advantage that the keeping of poultry has over many other hobbies (and it sometimes can be a disadvantage) and that is that poultry has to be attended to seven days a week—not forgetting Christmas Day too. In other words it is a continual interest and one which can be of great benefit to those who have previously thought of nothing else but work and adequately filling the time of those who have recently retired.

Potential Exhibitors
Throughout the length and breadth of the English countryside the traditional agricultural show is held annually, be it one of the large county shows, a town show or just a village show. At most of these shows there is a poultry section and whether it is organised by a special poultry committee of a county show or indeed the local poultry club at the village event, nevertheless, the interest and help is always there. It is at these shows, that countless thousands visit, that the various poultry breeds can be seen, and indeed where the advice that the would-be keeper might want is usually readily available. Someone with a leaning to livestock and who has a competitive nature may find his appetite whetted to keep poultry the day such a show is visited. A particular breed could well be attractive and eventually influence the purchase of some stock. A potential exhibitor is born through his or her own competitive spirit. Eventually that person, like myself, will gain experience and in a few years time

9

will be able and willing to pass on information to others who may visit that same village event.

That Patch of Ground

'If only we could do something with that patch of ground' or 'it's time that patch of ground was sorted out'—these are two comments that are often passed about amongst families. Very often the patch in question is a piece of waste ground belonging to one's own property which for various reasons has never been used or for which no real use has been found. It may consequently have become a wilderness and, in some instances, a dump. These are the patches of ground that poultry thrive on, particularly if they are properly managed. These are the patches of ground that can provide an annual income instead of an annual expenditure. These are the patches of ground that can be turned into neat little units instead of eyesores and these are the patches of ground that can give pleasure to a family instead of criticism and squabbles.

2

ACCOMMODATION

Having decided what really is the attraction of keeping poultry and being quite satisfied that it is not a passing phase the first thing that must be thoroughly investigated is the accommodation.

The poultry keeper really has three choices: either to keep his birds on free range, in battery cages or else in a deep-litter house. Dependent upon the type of method that is used is the number of birds which can be kept. There is no doubt that the recommended method for beginners is that of keeping hens on free range, but before describing the reasons why this is so, keeping poultry in battery cages or deep-litter houses should be briefly described because, in the future, you may want to go on and keep poultry commercially. Should this be the case keeping birds in batteries or on deep litter is certainly recommended because the birds will then be kept in numbers of multiples of thousands rather than double figures.

Battery Cages
Battery cages are specially constructed for laying hens. They are made of metal and are usually in three tiers. Various types can be purchased from the one where all feeding and cleaning has to be done by hand to larger blocks which are usually fully automatic as far as cleaning, feeding and watering is concerned. The cage which normally houses two or three birds has a sloping floor and as the eggs are produced they roll to the front of the cage for collection. Whilst the initial financial layout is high the great advantage to any battery cage owner is that he can spot

immediately hens which are not producing eggs and remove them right away.

Deep Litter

Keeping poultry on deep litter is a different operation entirely and indeed text books have been written on the subject. Most deep-litter houses have an earth floor (although some have been known to have wooden floors) and on top usually a layer of lime and then litter made up of shavings, peat moss etc. A droppings pit where the birds perch is usually provided and so too are communal nesting boxes. It is recommended that these should be dark to prevent egg eating. The one thing that has to be watched with deep-litter houses is that the litter is always kept turned and dry otherwise it becomes stagnant, birds tend to become unhealthy and egg production, the purpose of the whole project, starts to fall. Whilst the initial outlay is not as much as for anyone contemplating setting up a battery unit, nevertheless, non-productive birds cannot be spotted as easily.

Both these subjects are vast and require studying in great depth and detail. It was never the intention here to recommend or write at length about them but simply to acknowledge the existence of these two systems. If, after a few years experience, you want to think about moving over to one of these systems there are several books available specifically dealing with both battery cages and deep litter, as well as a variety of weekend courses, lectures and demonstrations.

Free Range

Free range is the ideal way of keeping a limited number of hens from which it is intended either to provide eggs for the house, table birds or even show stock. They can be kept very well and easily in a fenced-off pen at the bottom of the garden or in the orchard and management is relatively simple. But before embarking on this type of accommodation, consideration should be given to the number of birds to be kept and in this respect one plan may be to seek ideas and suggestions from local fanciers.

It must be pointed out, though, that estimates from experienced fanciers do tend to be high and it is often a good idea at the beginning to aim lower than figures suggested (see page 31). No doubt some poultry keepers and fanciers will argue that they have kept twice or three times as many birds as you eventually decide to keep. Maybe they have, but it could be the case that their buildings and runs were substantially overcrowded. Too much stock on any area of ground makes the ground stagnant and sour. Holes appear where the birds have been constantly scratching and grass simply does not grow. The whole area soon becomes an eyesore rather than an impressive management unit. It must be stressed also that where there are too many birds on any patch of ground, diseases of various kinds are found and breed quickly.

Let us now take a look at what is essential for the beginner keeping poultry on free range, the building, its erection and the surrounding pen.

Buildings

Having decided how many birds can be accommodated you must next buy a suitable building and this virtually means a wooden cabin (Figure 1). Perhaps those who have the skill to use hammers, nails and saws can make a suitable cabin, but it is stressed that it is a good idea to take sketches and measurements of one already in existence which is felt to be suitable for require-ments. Let·it immediately be said that making patched-up buildings out of old tea chests, corrugated sheets, sections of hardboard etc is absolutely out for many reasons. If poultry are worth keeping surely they are worth providing with a good home? Let us consider, therefore, the main essentials of a cabin whether it is bought or whether it is home-made.

Draught Proofing
This is probably the main essential. When growers are accom-modated in a building that is subject to draughts they simply do

6-8 PULLET HOUSE.

not grow as well as they should. Where laying birds are concerned draughts often have the effect of reducing egg production. Birds simply do not like draughts and neither does the professional stockman because his birds tend to avoid draughty parts and consequently there is not the correct turnover of litter in the cabin. As the birds try to keep out of the draughts, the place in the cabin they constantly occupy may get stale.

Waterproofing
Water pouring through a cabin roof onto dripping wet hens with empty nest boxes is a cartoonist's dream but not a poultryman's. All cabins must be waterproof. If they are not the result

14

simply means more work and more expense for the poultry keeper. Water coming in through a cabin roof or side continually onto the same patch of litter will soon make it smell. This means that not only has the litter to be changed but that it is a breeding ground for disease. Secondly, water leaking in must pass over the wooden structure of the cabin, consequently causing timber to start to rot which in turn will mean a costly replacement. It is perfectly obvious, therefore, that it pays dividends to start your poultry keeping correctly by providing your stock with a sound building. When purchasing or building a cabin there are several other points to bear in mind. Most of them are obvious but it is surprising how often they tend to be neglected.

Light
The placement of windows in your building is of vital importance. Your cabin should have plenty of light because there will probably be times in the winter when because of inclement weather your stock may have to be kept inside for a few days, and without the right amount of light the birds simply will not be able to see to eat. In addition, however, do not forget summertime and hot days. If possible, purchase a cabin with more windows on one side than the other and erect it with the windowless side to the sun. It is surprising how much heat can be kept out of your cabin and subsequently off your stock in this way.

The Perch
The perch is something which needs close attention because after all it is the 'hen's bed'. For parent stock a removable perch of a size 2in by 1in is recommended. Again, no doubt some neighbours who keep poultry will say that they just use the branch of a tree or any old piece of wood which they can lay their hands on for a perch. It may be firmly fixed in the cabin and everything appears to be fine but this method can make for two real problems. Firstly it can cause crooked breast bones and secondly, being fixed, it is not possible to remove it to creosote or disinfect it—a perch is often the place where lice and red mite can be found in

abundance. As it is the hens' bed keep the perch clean and make it of a comfortable size. Under the perch should be the droppings board (Figure 2) which is designed to keep the hens' droppings

PERCH AND
DROPPING BOARD

off the floor of the house and which should also be cleaned regularly.

Nest Boxes

Wherever a laying flock is kept the correct number of nest boxes (Figure 3) must be provided and placed in the right position. About 4 nest boxes are needed for every 10 hens and the box should usually be about 1ft square. The nest boxes should be placed with the bottom of the box some 2ft above the ground so that the hen has not too far to jump up to get into the box. More

Plate 1 'That patch of ground': A useful little cabin with a well-felted roof, correctly erected on concrete blocks giving sufficient space for a terrier to get under. A cabin like this does not encourage vermin. Note the pen has a separate entrance should there be need to go in it. The cabin is easily accessible from the garden thus preventing the hen run becoming muddy through a worn path

Plate 2 The broody box: note the ventilation holes and well-felted roof

Plate 3 Interior of the box showing: an empty section with the wire netting floor as described on page 55; a constructed nest of 7 eggs for a bantam; a nest of 11 eggs for a standard bred bird. Also note the dates the broodies were placed on the eggs which are chalked on the underside of the roof

NEST BOX.

important, the hen cannot stand on the floor and have easy access to the box to peck and eat eggs laid by previous birds. Far too often nest boxes in the form of orange boxes are placed in the corner of the cabin on a couple of bricks with some hay in the bottom, and far too often hens bored with life simply wait for other hens to lay and immediately break and consume the egg. Once hens have become egg-eaters there is little that can be done to cure them. Another important point to consider is that there must be some good hay or straw in the bottom of the box otherwise hens which often stand up to lay will drop the egg onto a hard bottom and a cracked egg will be the result. It is often said, too, that the more comfortable a nest box is the more tempting it is for a hen to lay and this is something that can always be borne in mind.

Ventilation

Providing it is the aim to let your birds out most days and have access to good ground where they can roam and scratch, ventilation will not present any real problem. Admittedly it is of great importance in commercial units of battery houses and especially deep-litter cabins but, as far as the small poultry keeper is concerned, providing the top windows of his cabin open back some 3–4in this should suffice. Most 12ft by 8ft cabins have three windows along the bottom to give light at ground level and three along the top of each side of the cabin. This should provide adequate ventilation. After all it is only during the night and in the darkest days of winter that your stock will be indoors.

Another continual source of ventilation in any cabin is 'the pop hole'. This is a hole strategically placed in the corner of a cabin where the birds can enter and leave at their choice. The usual size is somewhere in the region of 9in width by 12in in height. A little slide should be fitted above it that can be dropped to seal the building at night so that vermin cannot enter.

When purchasing or making your cabin, always ensure that all parts are completely accessible for annual creosoting and disinfecting and that as many fittings as possible can be removed. Perches and nest boxes should always be of the type that can be easily removed for disinfecting because it is in these areas that mites breed. Perhaps for a few moments the advantages and disadvantages of purchasing a completely new cabin rather than a second-hand one can be discussed.

A New Cabin

Buying a new cabin is often the cheapest in the long run. If it is your intention to have stock for many years, seriously consider buying a completely new building. Fanciers often argue that the biggest disadvantage with buying a new building is the price but there are, nevertheless, many obvious advantages. For instance, there are several reputable makers who will provide you with brochures of their buildings prior to purchase. From these brochures you can see exactly the size and type of building which might appeal and all the various extras that can be fitted. Another important point to bear in mind is that the prices of these buildings often include delivery to the site. In other words, with a new building you can see what you are getting before you pay for it and often there is a time guarantee as well.

Second-Hand Cabins

However, there are plenty of second-hand cabins to be bought at farm sales up and down the country and fanciers will readily observe that the obvious advantage here is the price. However, there are many disadvantages as well in buying a second-hand one.

Firstly, a fancier buying a second-hand cabin probably does not know what has occupied it before. It is quite possible that it could have been filled with birds that have had some sort of disease and a great risk is taken if this is the case. The cabin should be completely scrubbed and disinfected immediately it is purchased otherwise disease could be transmitted to stock which has taken years to build up.

Another disadvantage with the purchase of a second-hand cabin is that normally it has to be dismantled and moved to its new site. Often while buildings may look perfectly sound when erected, weaknesses may show up in main timbers when dismantling is commenced. In addition, of course, it could be argued that there is a genuine extra cost here in transporting the second-hand purchase to the new site. In my experience, a second-hand cabin never goes up as easily as it comes down unless it is a relatively new one. There are usually various bits and pieces that have been patched or replaced and erection of this type of building sometimes proves to be quite a job. Again, and bearing in mind the need for the building to be absolutely waterproof, it can mean a complete refelting of the building—another added cost.

There have been, however, many purchases of second-hand buildings which have been perfectly sound and providing you know the person you are buying from and are satisfied with the condition of the cabin, then there is every possibility that you can make a good buy.

Making a Cabin

Before leaving the question of what type of building to have, let us consider the possibility of making one. This is obviously a serious consideration, particularly if you are a good carpenter, but it is very much a third and final choice. Even with the purchasing of the best buildings and the best birds and equipment mistakes are made. It is highly likely that the inexperienced poultry keeper may make one or two major mistakes should he decide to make his own cabin. When the price of timber and time is all reckoned up as well, I would strongly advise purchase if you possibly can.

Erection of the Cabin

Whichever type of cabin is purchased there is one golden rule— all parts must be thoroughly creosoted *before* it is erected. It could be many years before the cabin is moved, so the coat of creosote put on the underside of the floor will have to survive years of wear and all weather conditions.

Eventually you will be ready to erect your cabin. The most convenient place for it will probably already have been chosen, but perhaps it is worth bearing in mind that the entrance to the cabin should be as near the daily footpath as possible. Too often, cabins are placed in the far corner of a hen pen and this results in a well worn track being established. This causes the soft ground to be churned up during the wintertime, which stock will certainly not relish, and nor will the poultry keeper who has the extra daily task of cleaning his boots. Always, therefore, try and have the entrance to the cabin and, indeed hen pen in the most convenient spot.

The one and only rule when erecting a cabin is to ensure that it is placed high enough from the ground. The normal height is anywhere from 9in to 1ft. Place the cabin on some sound cement blocks or sleepers, ensuring that the floor is level, and the rest of the job will be simple. The prime reason for setting the cabin away from the ground is, of course, to prevent the wood from rotting. If the floor is placed directly onto the earth, in a short time the main spars will start to rot and consequently the main structure will be affected. In addition, raising off the ground makes the cabin virtually one hundred per cent proof against any vermin that might frequent the premises from time to time. The cabin on the ground is a much greater attraction to rats and in such conditions they thrive. If there *is* found to be trouble of this nature, a good terrier dog has easy access to any such unwanted visitors.

The only piece of equipment that is then left after erection of the building is what is often termed 'the step ladder'. This is a simple piece of wood with little bars across which leads from the pop-hole entrance of the cabin to the ground, and placement of these bars at some 6in apart helps the hens to go up and down it, particularly in wet weather.

The Pen

Pens of all shapes and sizes, made out of all types of equipment, placed at various heights, can be seen around the countryside. There is really no rule as to what height the pen surrounding a cabin should be as it is entirely dependent on the type of birds kept. So too, of course, is the size of the wire netting mesh. I would suggest that for ordinary commercial hens, be they layers or table birds, a 3in mesh would be satisfactory. However, if you plan to keep bantams, a smaller mesh would be required.

Before erecting a pen or pens, some 3in by 2in timber will be required for the posts. Generally a post of about 5 or 6ft in height is best. With bantams, of course, 4ft posts will often suffice, particularly for the heavy breeds. Sometimes it may be necessary to put a top over this type of pen to prevent the birds flying out. Normally they do not take off from ground level and fly over a pen of 3–4ft in height but use the cabin as their first landing stage and then fly over from it.

A SIMPLE GATE LATCH.

In addition to the posts, if you plan to have more than one pen, in order to move your birds onto fresh ground from time to time, some corrugated sheeting about 2ft high must be used to form the basis of a joining fence between the two pens, the remaining 3ft or so being covered with wire netting. Such a fence prevents cockerels fighting and also has the added advantages of providing extra shade and acting as a wind-break.

Bear in mind that the pen should last for years and consequently certain extra precautions should be taken. Not only should the bottom of the posts be thoroughly creosoted (it is a good idea to leave them standing in an open drum of creosote for several days before use) but also the prevention of vermin must be considered. It is highly advisable to dig a narrow 9in deep trench around the pen where you intend to place posts and netting then actually to put the bottom of the netting in this trench (Figure 5). In other words, if you are using a 5ft netting, then only 4ft 3in of it would be seen above the ground. By

24

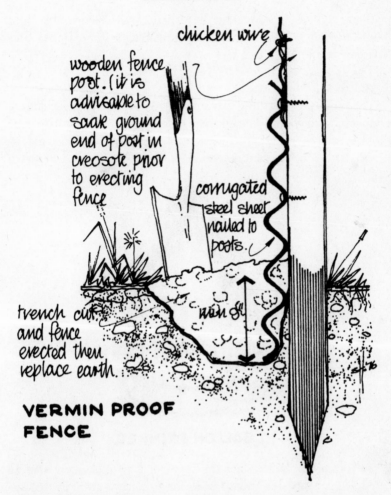

chicken wire

wooden fence post. (it is advisable to soak ground end of post in creosote prior to erecting fence

corrugated steel sheet nailed to posts.

trench cut and fence erected then replace earth.

VERMIN PROOF FENCE

adopting this method, vermin such as foxes, badgers and rats, which tend to burrow, are prevented from entering and perhaps causing a heavy financial loss.

Bits and Pieces

The picture is now virtually completed and the accommodation is rapidly reaching the point where it will be ready for the first stock. Suitable headquarters have been provided for the night and a good and safe run for the birds during the day but four items are still left unfinished.

Water

The most constant thing that stock require is water, and therefore an adequate water fountain must be provided. The standard type which can be found in all catalogues of poultry appliance manufacturers is highly recommended (Figure 6). One point to watch

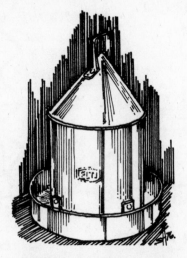

1 GALLON DRINKER

out for before purchasing the fountain is to make sure that all parts are easily accessible for cleaning, as it is strongly recommended that these fountains are cleaned at least once a month.

Food Troughs

Hens will also need something to eat from and a metal trough can be purchased from the same firm as the fountains. It is, of course, quite possible to make a wooden food trough, perhaps from waste pieces of timber, but the metal ones are less susceptible to harbouring disease and possibly last longer. Whilst the wooden ones are obviously cheaper they do, nevertheless, require scrubbing more regularly, particularly if a lot of wet mash is used. Poultry keepers can often be seen throwing food straight onto the floor of a cabin. Whilst this does help the litter to be turned over, it also means that the hens may pick up disease when they pick up this food and so it is advisable to use the food troughs at all times.

Grit Box

The third piece of equipment needed as far as the hens' diet is concerned is the grit box. This can be placed in the corner of the cabin because it is the piece of equipment which is used least. People are often under the misapprehension that, for the hen to make the shell of an egg, grit should always be provided. It is true that grit helps to ensure good shells, but the main reason for giving birds grit is to aid digestion.

Litter

And finally, what about the floor of the building? I would strongly recommend that this be covered with good clean shavings, possibly mixed with sawdust or peat moss to a thickness of 3–4in. Several handfuls of lime should be thrown onto the litter at irregular intervals as I find that this always helps to keep the mites down in the litter. The hens, too, pick up a certain amount of this lime when scratching in it and this will prevent lice and red mites making their home on the birds.

Sometimes it is noticeable that birds have their favourite scratching places in cabins and that, possibly in the odd corner, the litter does not seem to get turned quickly enough. A good tip is to fork all the litter into the centre of the cabin at least once a month and leave it there, letting the hens scratch through it and spread it around again. This keeps the litter moving and turned over and ensures that it keeps dry. Always remember that any wet litter must be removed as soon as possible.

3

BUYING STOCK

Once the hoped-for ideal accommodation has been erected the time has arrived for the purchasing of the first stock. However—be warned! This is probably the most important stage of all. If poor stock is bought then all the work that has been done is fruitless and it will cause nothing but worry, frustration and eventually a loss. On the other hand, if good stock is purchased sheer enjoyment will emerge.

Whatever happens don't be in a hurry to buy the first birds that are advertised for sale. In the natural anxiety to start keeping fowl a number of birds could be purchased that would lay the wrong colour of eggs or be too old and numerous other things. The best advice for anybody about to buy their first birds is 'be perfectly clear in your own mind exactly what is wanted'. It may take time to find the right birds and to buy them at the right price but patience in the long run will ensure the right birds are found.

Which Breed?
Obviously, the first thing to decide is whether large fowl or bantams are to be kept, and as, by this stage, pens have been constructed, no doubt this decision has already been taken. If you decide to keep large fowl perhaps the following points might be worth considering. The first point is whether birds are required simply to lay a good number of eggs for the house and for friends. On the other hand, you may have decided to keep table birds or may be even to purchase exhibition birds and start on a show career right away. Perhaps you have been influenced by some of the shows you have visited which encourage the

preservation of some of the rare breeds which were prevalent at the start of the twentieth century. As they are found in few hands today, you may feel that help can be given in preserving them. Perhaps none of these points cover your decision. It may be that a specific breed has been chosen because of its beauty. Perhaps a breed has been selected that lays brown eggs or even white eggs because you want to show the produce at local exhibitions. To help you make this decision the table given below may be of some guide and help:

Reason	*Breeds Recommended*
Good layers	Pure-bred Rhode Island Reds or White Leghorns and cross-bred birds ie Rhode Island Reds cross Black Leghorns or Rhode Island Reds cross White Leghorns etc
Table birds	Pure-bred Indian Game or Indian Game crossed with a heavy breed
Exhibition birds	Obviously there is a very wide choice here and most breeds have their respective classes at the bulk of the shows
A rare breed	One of the oldest breeds which deserves support is the Dorking. There are of course many others and the secretary of the very well organised Rare Breed Society would gladly advise on any breed which needs help and where reliable stock can be purchased
Breeds of beauty	Again, there are many beautiful breeds and possibly Hamburghs Pekin and Sebright Bantams are typical of birds that come under this heading

Reason	*Breeds Recommended*
Brown eggs/white eggs	If it is the colour of eggs which is sought after the choice is Marans for brown eggs and Minorcas for white eggs. These are the two breeds which normally lay the contrasting colour of eggs that take the prizes at the shows but it must be added that Welsummers can also be included in the brown egg brigade

Perhaps the smaller bird, the bantam, has attracted you. There are a good many reasons why this might be so. If you are short of space, bantams are an obvious choice. As already mentioned, the bantam is an ideal bird for children and the family to enjoy. It may well be that the show spirit is well and truly in your blood and you cannot wait to place stock in the show pen and see it win prizes. In this case, ready-made show bantams might well be your answer though if you do have the patience you can, of course, breed potential winners yourself provided that the correct breeding stock is bought to start with. Finally, despite their small size, bantams are very good value. For a smaller outlay than for the larger birds, and lower feeding costs, the eggs produced are comparatively large. In other words, the small bantam is more cost-effective than the larger types of fowl.

Where From?

Looking down the chart of the large birds obviously a commercial breeder is the man who will provide beginners with laying or table birds or birds to lay brown eggs or white eggs. However, a specialist breeder of show stock is the man to contact if exhibition birds are required and he, too, can often help as far as rare breeds are concerned. Needless to say, showmen can also help with birds for laying different colours of eggs.

As far as bantams are concerned it is really the showmen who must be located. Obviously contact will have to be made with him if show birds are required or breeding stock to produce them and so, too, if bantams are wanted for children's pleasure. He is the best man to supply these because not every bird bred by a specialist is correct—very often breeders are only too happy to sell off some of their mismarked birds to children. But again be warned! The soundest advice I can give is to recommend you to go to a well established and reputable breeder and tell him exactly what you want. Describe the breed, the age, the stock and above all tell him exactly what the birds are intended for because, obviously, there is a big difference in the price of birds that have a chance of winning a major show and those which are simply required for beauty and running round an orchard or back garden. If the breeder has not got what is required he will in most cases give another name and address. If you have any difficulty in locating the right stock then seek advice from the Poultry Club of Great Britain, (the secretary's name and address can be found at a library or information bureau) where willing help will be given. Normally, the secretary will happily give the name and address of breeders in any area and in addition will supply the name and address of the breed club secretary so that if local breeders cannot help at least a breed club secretary can give a lending hand to sort out the problem. Always remember personal recommendation is the nearest and quickest guarantee of success.

How Many?
Mistakes are often made by inexperienced keepers buying too many birds when a small unit is being started. If poultry has never previously been kept it is so wrong to do this but in everyone's eagerness to fill their cabins it happens time and time again. The old saying, 'walk before you can run' surely applies. Start poultry keeping in a small way, particularly if you have never kept birds before. There is so much that can go wrong that if you start with only four or six birds the possible loss is nowhere near as great as if you have a cabin-full—finance must, of course,

31

always be kept in mind. Go along slowly gaining experience day by day.

Perhaps the ideal number for anyone to start with is one male and three females. This probably sounds far too few—far less than what was envisaged but think of the reasons already mentioned. If something should go wrong with four birds they can easily be replaced. If something happens to forty-four, you probably won't want to hear the word 'poultry' again! Gradually build up the stock and when you have become more confident in what you are doing and, in particular, have experienced summer and winter conditions, then you can start to increase the numbers you keep.

Age

All sorts of different breeds at all ages are offered at different prices. As is often the case, it is not always the cheapest that is best. Obviously, much depends on whether it has been decided to keep a few birds purely for laying purposes or whether birds are being purchased to be the foundation of potential show birds. If you want laying birds to keep the family supplied with fresh eggs the soundest advice is that the birds should be of the current year's breeding ie point of lay pullets. It is at the age of five to six months that a bird is ready to start laying its first eggs and, small though they may be at first, they soon increase in size. A hen always lays more eggs in its first year than any subsequent year and the older it becomes the fewer eggs it lays annually. Therefore, if you buy point of lay pullets then, obviously, the best return both in financial terms and egg output can be obtained from them.

Then comes the question of the male bird. Again the first thing to do is to make sure that he is from a completely different strain. (This does not necessarily mean that he should be from a different breeder because many breeders carry several strains of the same type of bird.) The reason for choosing different strains is obviously fertility. If a male bird is too closely related to his hens then fertility will certainly not be as good and, secondly, chickens may be born with various deformities.

If it is show stock from which you intend to breed, then age really does not matter. What does matter is that whatever birds are bought come from a reputable breeder, and before buying them have a look at some of the stock they have bred if at all possible. This will then give you a good idea of the type of birds you can expect to get from the poultry you are considering buying. Providing your show stock is three years old or under the female birds should lay sufficient eggs and the male birds be fertile.

Many breeders will strongly recommend that females are purchased from one breeder and the male bird from another. Sound advice. Many successful birds have been bred this way but as mentioned above breeders do keep different blood lines of birds and if the same blood can be obtained, not too closely related, from a stud producing consistent prize winners, this may well be the right answer. When buying birds it is always best to make a personal visit to see the stock. Never buy birds from anyone who keeps them in dirty and unhealthy conditions because besides buying birds, you will be buying trouble as well. Again, when buying birds always ask the breeder what he uses for feeding them and if this differs from what you believe to be correct, take his advice and use his type of food until the birds become settled into their new surroundings. As far as changing them onto your own particular food is concerned, do it gradually. Mix some of it with the type the breeder was using and then gradually add more of yours and less of his until the birds are firmly established on the type of food on which it is intended to feed them.

The New Home

When the birds eventually arrive in their new surroundings another piece of advice is to give them a spray or dust around the vent and under the wings with a recognised lice and red mite powder or spray. This will ensure that any mites that they may have brought with them are not transferred to their new home. The other most important point is that for the first few days they should be kept inside their cabin. Don't be showing all the friends and neighbours the new stock of which the family is obviously so proud. Leave the birds alone in their new surroundings, as quiet as possible, well fed and, particularly, well watered so that they can get used to them. Another reason why the birds should be kept in is that if they are given the freedom of their new home, when darkness falls, they will not all return to their cabin as they won't yet know their way around. The correct time to let out new stock into their various pens for the first time is in the late afternoon before giving them the evening feed. Let them out on a dry day and as the evening is drawing in some corn thrown into a trough will attract them back inside the cabin again.

Plate 4 The corner of a building with four chick runs erected by the use of concrete blocks. Note the removable frames on top and the labels giving reference to the pen the chickens were bred from and the hatching date. Also note the simplicity and tidiness; as the chickens grow these runs are extended by the addition of more blocks

Plate 5 A hay box with strong netting on top and wire netting running all around the bottom to give maximum protection from such vermin as foxes

Plate 6 Chicks which have been transferred to their second home and are now able to exist without artificial heat. Here the breeder is using a section of a cabin rather than a hay box or night ark. Note the food trough, clean shavings, natural light, water fountain and egg tray which is used to encourage the chickens to eat and prevents the wastage of food

4
CARE

By now a great deal of time, effort and money has been spent in establishing a small poultry unit. It is, of course, absolutely no use having the best equipment and the best birds if one does not know how to look after them correctly. It is true to say that there are fanciers who simply throw their birds mixed grain once a day, top up the water fountains and the birds from then on virtually look after themselves. Indeed, what is most surprising is that they do lay a few eggs! This, of course, is not the recommended way to care for stock. What is?

Care really means in simple terms 'management'. To manage a poultry unit, whatever the size and whatever the type of bird be they large fowl, bantams or waterfowl, is similar to managing any business or section of it. A correct routine must be found for feeding and cleaning and, above all, the poultryman (or manager) must have an eye to spot any weakness.

There are three basic requirements of a good stockman and without them, trouble will almost certainly follow. These requirements apply to any stockman looking after poultry whether he owns four, forty-four or a complete commercial battery unit! No bird can live without food and water; no bird can give of its best if it is not kept in clean and healthy surroundings continually maintained to a high standard. In short, the three basic requirements are: feeding, cleaning and maintenance.

Feeding

Generally speaking, birds are fed twice per day although there are many fanciers who because of working long hours and shift work can only feed their birds once a day. For those who can feed twice a day the normal routine whether feeding layers, table birds or bantams is to feed a damp mash in a morning. The prime reason for this is that it is the bird's main meal of the day and it is generally accepted that it is the meal that does them the most good. A recognised brand of mash is full of vitamins and proteins so necessary to produce eggs and to keep the birds in good condition. The mash must be mixed with warm water until it is 'damp'. A wet slushy mix is of no use at all and will tend to be ignored by your stock. Obviously, a 'layers' mash is given to laying birds and a 'fatteners' one to table birds.

In the evening the normal feed is to give laying birds mixed corn. This just tops up the crops for the night and by the time the birds alight from the perch at dawn the following morning their crops are empty and ready to feed again. In the case of table poultry a brand of pellets specially manufactured should be fed to the birds at night.

But what about the poultryman who can only feed once per day? He will normally feed his mash and will in many cases leave a hopper in the cabin filled with dry mash which the birds can continue to peck after the wet mash has been consumed. The great advantage of this, besides giving the birds the food they should have, is that it prevents boredom and keeps your poultry out of trouble. If a bird is hungry it will continue to look for food and consequently will not get involved in such vices as egg eating, feather pecking etc. Some fanciers who can only feed once a day do give their birds mixed corn (known in Canada as 'scratch'). Whilst this does satisfy the birds it does not necessarily give the return that is required. It certainly fills the crops up but fills them to the extent that they are fully satisfied and will then tend to pick up the bad habits listed above. Also, it is generally accepted that mixed corn does not have the protein and vitamins necessary to give the best egg production.

Scraps
I mentioned in an earlier chapter that one of the reasons for keeping poultry was to use up the scraps from the kitchen.

Unfortunately too often such scraps as stumps from cabbages and cauliflowers are simply thrown into a pen and the birds eat only what they desire leaving the remainder to rot and subsequently smell. Try hanging a string from the apex of your building to a point 6in above the litter and tie scraps such as cabbage stumps etc onto this. Once the birds get used to it they will peck at this string for hours on end and so keep from getting into any other bad habits that might tempt them.

Another mistake so often made is for all scraps to be put on one side then mixed into the mash for feeding to the hens. This method of administering the scraps is correct but it is wrong to feed *all* household scraps. It is generally accepted that the odd crust and waste potatoes, vegetables, perhaps stale biscuits and cake are permitted, but it is advisable to avoid feeding potato peelings and all forms of meat to your poultry. Far too often peelings are mixed into the mash and the birds certainly will not eat all of them so that, like the cabbage stalks, they tend to lie about the unit until they rot, as would meat scraps such as the carcass from the Sunday joint. A good policy to adopt is only to mix into the mash the type of food waste you would not mind eating yourself. If you do this you will generally find that your birds will leave very few scraps around. In fact, mixed with the mash, such scraps do them a power of good.

Mash is usually mixed in a galvanised or plastic bucket and after it has been used several times a lining tends to become stuck onto the sides. There are many fanciers who continue to mix their mashes day in and day out in the same bucket, quite content to let the lining get thicker and thicker. Obviously this should not be done because when the annual scrape is given to the bucket it will be found that the mash sticking to the side of the bucket is dry, crumbly and stale, and of a greenish colour. Always wash utensils out after they have been used. After all, and no apology is given for making this statement, birds are only like human beings. Any normal person would never dream of eating off the same plate for 365 days per year without washing it up after each meal. It is only the same with poultry.

In addition to their foods, birds should be provided with a form of grit to help digestion. Any corn merchant will recommend the best type. Some fanciers, however, never feed their birds grit and the birds don't seem to suffer any ill effect.

How Much Food?

A common mistake for beginners is to give too much food in the initial stages. There are no hard or fast rules as to how much a bird should eat and it does of course depend on whether large fowl or bantams are being kept. However, as a guide, in the case of large fowl enjoying free range, it is generally recommended that 2oz per bird of layers' mash be fed in the morning and then 2oz per bird of mixed corn given in the evening. For bantams these amounts can be halved.

Every fancier must take into consideration the amount of natural food that the birds will eat as they fend for themselves in their respective runs and from their pickings along the hedgerows. A good guide and something that should always be borne in mind is the size of the bird's crop. Anyone keeping poultry for the first time should catch one or two of his birds after feeding and feel just how full the crops are. Obviously if they are not full (and they should never be crammed) a little more food can be given at the next feed and the correct amounts will soon be recognised as experience is gained.

Cleaning

The cleanliness of the premises is something that a good stockman always has in mind. If disease is to be prevented it is of prime importance that the cabins especially and the runs are kept well up to standard. Generally the cabins should be cleaned out weekly but this does not mean an entire spring clean throughout. Providing the perches are scraped, the dropping boards cleaned and fresh sawdust scattered over them, everything should go smoothly. It is not necessary to remove all the floor litter weekly. Providing the hens are continually turning this and the litter is dry then in most cases it can last for many weeks. Shavings and sawdust are acceptable litter throughout poultry houses, although many older fanciers tend to use lime especially on the dropping boards.

A good stockman, when cleaning his cabins, should also check his perches. Periodically, they should be painted with creosote along the underpart though not on top if the cabin is being occupied. Again, the reason for this is to keep mites away where they are most prevalent. If the perches were by mistake creosoted on the top as well as underneath it would not only mark the feathers of the birds but also could tend to burn the breast. Nest boxes, too, should be checked weekly to ensure that there is adequate straw or hay on which the hens can safely lay their eggs, and given a periodic spray of a recognised mite deterrent.

After cleaning has been completed what about the manure? It is a common mistake for a pile of manure to be tipped in a corner of a hen pen and left there to rot. This soon develops into a running and horrible mess and of course gets stagnant. Birds will frequent it and if there is any disease in the residue it is the quickest way to pass it on to the rest of the flock. It is therefore strongly suggested that all poultry manures should be kept entirely separate and completely out of the way of the occupants of the cabins and runs.

Maintenance

Maintenance of a cabin and poultry run is similar to the periodic maintenance of any house or building. Checks should regularly be made to ensure that the roofing felt is not torn or cracked and that all windows are in place. It is important that when a poultry house is empty it should be given a thorough spring clean and this means painting it both inside and outside with creosote. This not only preserves the wood but, as already mentioned, eliminates any mites or lice that find their hiding places in various odd corners.

And what about those other jobs that must be done to keep the unit efficient? Much emphasis has been laid on the control of lice and mites because if this is not done birds will soon start to suffer. In addition to the methods already recommended, it is suggested that at least every couple of months, as the birds have taken to their perches in the evening, they should be lifted off and given a dusting with a recognised powder or spray which has the effect of eliminating these unwanted insects. There is nothing more annoying to a bird than to be infested with mites and it is surprising how these have an effect on the bird's health and consequently egg production. The best preventative is to put nicotine sulphate on their legs. All that is required is a feather dipped into the nicotine sulphate bottle then just a touch applied around each leg. This method is long-lasting and will certainly stop anything which tries to crawl up the hen's legs and eventually onto its body. Do bear in mind, however, that nicotine sulphate is a poison and must be locked away out of the reach of birds and young children.

Perhaps at this stage it is advisable to list a few of the basic ailments and vices of poultry that might confront someone keeping poultry for the first time. It is not intended to give a complete rundown of all the diseases of poultry but the opportunity should be taken at least to mention some of the basic ailments.

Bronchitis and Colds

These are usually caused by draughts in the cabin and exposure to extreme weather conditions. The symptoms of these diseases are for the birds to have running eyes and noses, to be heard to sneeze and to have what is often termed a 'croak in the throat'—a rattling sound coming from the throat.

Bumble Foot

Bumble foot is caused by birds alighting from perches and dropping boards which are too high from the ground onto a hard surface. It is therefore important not to have the dropping board more than 2ft above the ground and for the floors of the cabins to be covered with at least 3–4in of litter. Bumble foot in poultry is similar to a corn on the foot of a human being.

Coccidiosis
This is a disease which effects young stock, particularly young chickens between the ages of two weeks and four months. It is a parasite which affects the intestines of the birds. If a batch of young birds look weary, wings down and lethargic then it is a sure sign that coccidiosis is present and examination of the droppings will generally show that blood is present amongst them.

Cropbound
The cause of a bird becoming cropbound, in nine cases out of ten, is because of overfeeding. There are birds who will just simply eat and eat until they cannot digest any more and this is when the trouble starts. One remedy often used for birds that are cropbound is to dig up some worms from the garden, push them down the affected bird's throat, and usually they will work their way along the inside of the crop and clear the trouble away.

Fowl Pest
This is a disease which every fancier will certainly have heard of at some time or another and it is basically pneumonia in poultry. This is best recognised by birds losing their appetite, rattling sounds in their throats, droppings of a greenish colour and the birds being entirely mopey, with a continual nervous movement of the head. If they are laying birds it will be noted that egg production will become almost non-existent and if it is anyone's misfortune to suffer this disease the local Ministry of Agriculture office must be notified immediately.

Worms
Like all other birds and animals, poultry do from time to time suffer from worms. This is not a disease that happens every month or every year and flocks of poultry can be kept on the same

ground and in the same houses for years without having this trouble. The presence is not easily recognised because the birds continue to eat reasonably well and to lay and only by constant handling can the presence of worms be detected. The usual thing is for a bird to start losing weight and to get so thin that its breast bone becomes like a razor even though its crop can be found to be full.

In the old days, many weird and wonderful remedies were put forward to cure poultry diseases and indeed some of them are still used today with great success. However, with the tremendous progress that has been made through detailed research in modern medicine, quick and reliable cures are to be found. For any beginner who finds himself struck by disease his easiest and best way to cure it is to seek advice from his local veterinary surgeon.

But there are other things that poultry suffer from, which, whilst not diseases, are certainly vices and none more so than egg eating and feather pecking.

Egg Eating

Egg eating is usually started by boredom amongst the birds. If they are shut up with nothing to occupy them then they are certain at some stage to take an interest in the newly laid eggs in the nest and when the first egg has been cracked open others will surely follow. The main culprits can always be identified by the stain of the yolk on their feathers. By far the best and easiest way to cure an outbreak of egg eating is to provide a communal nest box with a small side entrance so that the hens when entering it actually enter into a very dark place. Once eggs are laid, in the majority of cases, the hens will not be tempted to search for them. If this does not work, another remedy is to take a fresh egg, blow it and then refill it with mustard with the ends of course securely sealed. If the offenders attempt to eat these eggs it is usually the first and last time that they try.

There have been times when both these suggested remedies have failed, and the only avenue left open is to debeak the offender. It should be stressed that only the top beak should be cut and, before attempting to do this, advice should be sought from an experienced stockman. There is only a recognised amount of beak that can be cut otherwise the health of the bird will be affected through its inability to eat sufficient food.

Feather Pecking

This can be found in young birds from four weeks old to maturity. There is nothing worse than to see a fully grown bird with just a small stump for a tail, and the remedy lies in the hands of the owner. Some fanciers believe that it is due to some vital ingredient missing in the food but, in my experience, the real cause is either the presence of mites or boredom, particularly the latter. If birds are properly looked after and can continually occupy themselves scratching and seeking food in their runs then this habit is rarely seen. Feather pecking is something that I personally have never encountered simply because my birds have always had plenty of free range and plenty to occupy them even when shut in their cabins in bad weather.

Utensils

Amongst the other odd jobs which occur from time to time is the cleanliness of the equipment being used, particularly the water fountains. No matter how much clean water is used the fountains on the inside will gradually become covered with a thin film of green slime so water fountains should be scalded out and scrubbed at least once a month.

Vermin

Another check to make regularly is to ensure that vermin such as mice and rats are not present. Both these creatures can do a great deal of damage in a poultry house. Mice are most expensive creatures when they start to nibble the food bags and rats, of course, can be a constant menace stealing eggs and young chickens as well as helping themselves to a fair share of food. There is no need to go into great detail as to how these two pests should be eliminated—sufficient to say the sooner the better.

It is rare to find vermin if your cabins and pens have been built in the manner recommended. However, if an unwanted intruder does find his way into your pen, and particularly if it is a fox, there is just one way in which it is possible (although I cannot guarantee it!) to prevent it from entering your cabin. This is by the erection of a fox tunnel (Figure 7). A simple diagram is

given above. From the pop hole the fox has to go along a 5ft tunnel which is made by providing a 1ft 3in high board set about 1ft from the side of the cabin. In addition before the fox gets in the tunnel a chain is simply hung down in front of the pophole and it is believed that when the fox sees the chain and the long tunnel it thinks it is a trap and decides not to enter. It is pure psychology but it has been known to work.

Are you or will you be a good stockman?

5

BREEDING

It is only natural for any poultry keeper who has established himself over a minimum period of twelve months to start thinking about the possibility of changing some of his stock. During the preceding months much experience will have been gained as a stockman and, more important, contact will have been made with other poultry keepers in the area. Discussions will doubtless have taken place on what should be done and much advice given—albeit some of it not necessarily sound! It should be emphasised that only advice from a person who has proved himself and his birds to the utmost in confidence and ability should be accepted. Many different ways of rearing chickens have probably been the subject of much 'garden wall' gossip and, quite probably, the newcomer is left in an utterly confused state of mind. Basically, what should be considered as far as replacement stock is concerned is whether it should be bred or bought and if it has to be bred—what does the poultryman want to breed?

Parent Stock

The first thing to be taken into consideration is the question of parent stock. Has the poultry keeper been satisfied with his own birds to date and would he be happy if he bred replacements exactly the same? Obviously this differs as to whether breeding is to take place purely for egg laying or for exhibiting. Should the decision have to be made over a laying strain then the choice is easy. The birds have either produced sufficient eggs or insufficient and that question is easily answered. However, as far as exhibiting is concerned, this is the subject of another question that requires much thought and very detailed answers.

Breeding for showing is a very complex subject and whole chapters have been devoted in other books to this matter. Much depends on the breed kept—in some breeds the emphasis is on shape, in others colour, in others markings. It may be a question of using the cock of a deeper colour with a hen of a lighter colour. In several breeds of poultry the practice of double mating is used. Without going into great detail, this means that the showman has pens from which he breeds his male birds to the standard laid down by the Poultry Club and separate pens where he breeds his females to the recognised standard.

On the other hand, it may be table birds which interest the fancier and the question to be answered here is—have the birds made the weight which was wanted and made it in the period of time they should have? The answer is ready-made and the fancier will then have to decide whether to breed from existing stock or bring in new birds.

The Breeding Pen
Whether breeding for table birds or for laying birds the number of birds required in the breeding pen is the same. So too is the number which many of the fanciers who breed purely for show needs. Apart from the marked breeds which, in some cases, necessitate double mating as explained above, many of the country's champions have been bred from pens of one male bird and six females. It must be stressed that to breed show winners you don't always breed from just one male and one female completely separated from all other stock on your premises. The ideal number of birds for all these categories is one male and six females and perhaps a look should be taken at the male and females concerned.

The first priority as far as the male bird is concerned is that he should be fertile otherwise the whole thing might as well never have been started. The next question to consider with both male and females is that they should be really fit and healthy. Breeding never, ever should be started with birds that are not 100 per cent healthy. As far as the male bird is concerned, age should be taken into account—in other words will he last for the entire breeding season? It is strongly suggested that when male birds reach the age of three or four, this should really be the last year that they are used. This, however, does not necessarily apply for someone who is breeding show birds. It could well be that a certain male bird, even exceeding the age of three or four, which has previously produced magnificent stock can still be used. Finally, the other main point to bear in mind is that the male bird conforms to standard whether you are breeding birds for any of the three categories listed.

And what about the females? Again they must be true to the breed in character and standard whether utility birds or pure-bred birds and they must be healthy. Their egg-laying potential must seriously be considered and so too must their age. It is admitted that many people have bred successfully from pullets (that is a first year bird) but this is a practice which is not necessarily generally recommended. Every fancier certainly requires good stamina in his chickens and it is suggested in many quarters that females from two years old upwards only should be used. It is not always possible to keep to this rule but it has been illustrated in the past many times over, that good robust chickens can usually be bred from birds of this age.

It cannot be stressed enough that the making up of any breeding pen for any purpose whatsoever is one of the most important decisions in the fancier's year and particularly as far as the male bird is concerned. If the wrong bird is chosen then there is a very good chance that all the chickens bred could be wrong in some way or other no matter what the females are like. In six females, if one or even two bad decisions are made, there is still the likelihood of getting some excellent stock from the other four.

The next step is to make sure that the male bird has no sharp spurs. Spurs start to grow on cockerels from 12 months old onwards and at the age of two or three, in some breeds, can become at least 1in in length and very sharp. Fanciers using male birds with spurs of this nature can run into trouble for the hens will often fight shy of the male bird for fear of being torn) and consequently, infertility will abound. The spurs should be trimmed but care must be exercised. They must not be cut off completely otherwise bleeding may take place and often the bird can become lame. A pair of pliers should be used, roughly one quarter of the spur's length should be snipped off and the end just rubbed down with sandpaper to make it smooth. In other words, it is like a human cutting finger nails. As far as the hens are concerned, clipping of the feathers around the vent is a practice which is often successfully used to improve fertility, and apart from this all that is required is a jolly good dusting to ensure comfort for the birds at the start of the breeding season.

Care of Hatching Eggs

Fanciers commencing keeping poultry and breeding for the first time are often very mixed up on the care of the hatching eggs and indeed how long they should have the birds together before eggs can be considered fertile. Without doubt a breeding pen should be together and the hens in lay for at least ten days before any eggs are collected for hatching purposes. Similarly, should it be necessary to change the male bird during the breeding season ten days should elapse before eggs are collected from the new male bird. What, then, should be done with the hatching eggs when they are collected and how long can they be stored for?

An egg will last for a maximum of 14 days before it is put in an incubator or under a broody hen, providing and only providing that it is kept correctly. The best way to store the eggs are on the usual type of egg trays with the point of the egg downwards and they should be stored in no more than room temperature—they should never be put in cold cellars or warm greenhouses. Each day the fancier should simply turn the eggs in the tray, and by the time that the fourteenth day has arrived a good number of eggs will have been collected. Providing the male bird has done all that has been expected from him, then given correct hatching procedures, chicks should arrive.

Many new fanciers hear and dream up all sorts of tales that eggs should be left in the nest until the birds have laid sufficient to sit on or that they should be stored in warm ovens etc. This is simply not the case and providing the method described is followed in detail then the fancier should be trouble-free and well on the way to getting his first chickens.

The next question to consider is the hatching method. It is certainly no use having a clutch of eggs ready for sitting and the fancier scratching his head wondering how and what he is going to do. Time will run out and disappointment will prevail. There are basically three methods of hatching chickens and any new fancier would do best to try them in the order given. Firstly, there is the good old-fashioned method as far as small poultry-keepers, and indeed many successful exhibitors are concerned—the use of broody hens. Secondly, there is the method for the more experienced fancier and one who keeps a good number of birds—the use of his own incubator and thirdly, there is the method of custom hatching. This is a method where eggs are taken to a recognised hatchery, where they are incubated along with countless hundreds and possibly thousands of others. The fancier either pays for the number of eggs put in or, in some cases, the number of chickens eventually produced. But for this exercise, the first method will be discussed in detail and some information given about the second one because these are usually the methods that the small poultry keeper and indeed the well known and experienced showmen adopt.

Broody Hens

What is a broody hen? A broody hen is a bird which has laid a batch of eggs and where nature has played its part by persuading the bird to sit—in others words to go broody. Broodiness is also encouraged by warm conditions and consequently, broody hens in winter are hard to come by. In fact, a broody bird in November, December and January is a jewel as far as showmen are

Plate 7 A compact bantam unit situated at the bottom of a small garden

Plates 8 and 9 The two stages in washing a bird. See page 86

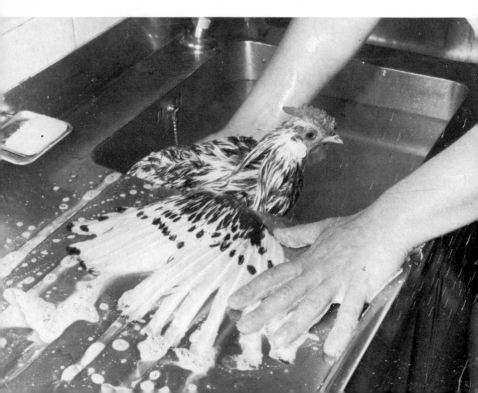

concerned. A good broody during these months means the showmen can incubate eggs which will produce what is termed 'early chickens' and which can be used for the mid-summer shows. These birds often score over the older ones that have started or are even in the middle of moulting.

How does one make hens broody and where can a broody hen be found? This again is a question not easily answered because one of the most difficult things to persuade or alter is nature and very often it is very difficult and dangerous to interfere with it. The old phrase, 'let nature take its course' is often sound advice. However, during the summer months it is possible in some instances to persuade birds to go broody by leaving eggs in a nest box.

Having decided that the eggs should be hatched by a broody hen what is the correct procedure to adopt? It is perfectly true that there are many poultry keepers who simply make a nest in an old orange box or tea chest and place the broody hen on it and leave it to its own free will. Often two or three such nests can be seen in the same building and apart from throwing the broody hen some corn the poultry keeper pays little attention to them until hatching is near. It is further true that some of these methods do work and indeed many chickens have hatched this way but it is suggested that many more would have hatched if the nest had been prepared properly.

The ideal way to make a nest for a broody hen is to build a special broody box as shown on page 18. These boxes normally have a sloping roof and two, three or even four compartments of a size 12in in width, 15in in length and some 15–18in in depth. It is essential that the roof is well felted to prevent any rain leaking in during the period of incubation. However, the most important part is what is to go on the bottom. On no account must boards be put on—just good strong narrow-guage wire netting. This not only prevents vermin entering but it is also the key to success—the box is placed on the earth and the heat of the hen's body draws the moisture from the ground and consequently prevents the eggs from drying up. It goes without question that this is the method much preferred by many fanciers and hatching results have certainly proved it is a method to be recommended.

The next problem is to shape the nest, and for this purpose some clean straw is required. A handful should be placed in the necessary compartment and made to give it a 'basin' shape. Furthermore, some pot eggs or eggs that are not intended to be used for hatching should be placed in the nest. All is then ready for the arrival of the broody for her three-week stay if she is hatching hen or bantam eggs, extended to four weeks for geese, turkey and duck eggs. The best time to place the broody on the eggs is in the evening as by setting her at this time she does tend to settle better. For the first 24 hours she will put the finishing touches to the shape of the nest by wriggling around in order to secure her own comfort. Once settled, the false eggs should be taken away and the ones intended for hatching placed underneath her. Before she starts on her marathon incubation she should be given a good dusting with a recognised lice or mite powder and then all should be well.

It goes without saying that broody boxes should be placed in a quiet area and near a small cabin in which there are no livestock. This will enable the poultry keeper to use it when he takes the broody off for her once-daily feed of mixed grain. Admittedly, broodies can be taken from the nest both morning and evening, but once per day is generally sufficient and the reason for giving mixed corn is that the hen will fill her crop quicker preventing too long a delay before returning to her important clutch of eggs. Broodies should be let off in the evening when the fancier probably has more time on his hands should anything be wrong. Needless to say, fresh drinking water should be available where the broody is fed.

After the first fourteen days the eggs can be tested for fertility. Again the work of the stockman takes place at night as he takes the eggs one by one from under the bird. A strong flash-light with a good beam is all that is required. If he sets the egg between his index finger and thumb, with the flash-light underneath, the egg can be 'candled' to test for fertility. Eggs that are completely clear are obviously infertile and those which show a huge dark patch with just a small clear end in the thick end of the egg are the ones in which there are chickens (Figure 8). The infertile eggs can

FERTILE EGG.

then be removed from the nest and instead of the broody sitting on her ten to twelve eggs (much depends on the hen's size) she could well have eight for her remaining seven days. It is then that the work of the poultry fancier really takes place—helping the broody to hatch the chickens.

How can the fancier help a bird sitting under these conditions to hatch chickens? It has already been agreed that a certain amount of moisture can be drawn from the ground and as the egg shells are porous it will help to prevent them baking and becoming so hard that the small chick inside cannot crack them to get out. Three or four days before hatching it is recommended that some warm water be splashed onto the nest whilst the broody is off for feeding. Some fanciers damp their nests very thoroughly especially when they are using good reliable broodies and by doing this all the assistance that is required will be given to the hen. On the 20th and 21st days, when sitting hen eggs or bantam eggs, the chickens will gradually start to chip the shells and eventually work their way out. No real help should be given to the chicks in order to get them out of their shells. As in all birds there are weak chickens that just have not got the energy nor the formation to get out of the shell and if hatched can become an embarrassment to the fancier. Although, of course, there is no harm in peeling back a little of the shell to help the chick on the way there is a belief among many poultry fanciers that if a chicken cannot get out of the egg on its own then it really is not worth bothering about. This is generally speaking quite true.

But what about those poultry keepers who have for many years simply put their broody hens on straw in some old orange box lying in a corner of a shed? As mentioned previously, chickens can hatch under these conditions but the great danger is that the whole lot could dry up through the nest not having any real moisture. If you do keep poultry, keep them under the ideal conditions and take up recognised ways that have been proved over the years. The method of sitting broody hens in the ground is a very old-fashioned one but if you consider how birds in the wild hatch their eggs, you can see how true to nature it is.

Incubator

And what about the fanciers who decide to use a small incubator for their first hatch of chickens? There is nothing against this method providing the fancier sticks rigidly to the instructions laid down by the incubator manufacturers and temperature and moisture are the two most important items which must be controlled at all times. If these are wrong the eggs will bake and chickens will be lost. It should also be mentioned that whereas a broody hen automatically turns the eggs over as she moves around the nest, the eggs in an incubator must be turned by hand (unless the incubator is automatic) twice per day. This means that if the poultry keeper is going to be away for any length of time then arrangements must be made for someone else to turn the eggs and check the moisture and temperature. Possibly one advantage that fanciers using an incubator have over those using a broody is the fact that they have the excitement of seeing their chickens hatch but after the first two or three times the attraction of this becomes less and less and, generally speaking, incubator chicks are left alone in the same quiet way as those under a broody hen. There is of course the consideration of cost to be taken into account. Incubators cost money, whether run on oil or electricity, whereas broodies cost very little.

Custom Hatching

For those fanciers who are lucky enough to have a commercial poultry breeder in their area, who practises custom hatching then this method of hatching eggs can be organised quite successfully. The most important and indeed only item to be carefully checked in this respect is of clearly defining eggs which belong to you. A simple pencil mark with an initial, or a particular pen marking is all that is required, together with a careful date made of the time the eggs are due to hatch so that the chickens can be collected at the precise time by arrangement with the poultry breeder. On collection, of course, an adequate container must be provided for taking the chickens from the hatchery to their first home. A cardboard chicken box could be purchased for this particular job or alternatively a good strong box, not too deep, with some clean hay in the bottom to keep the chickens warm, is quite sufficient. Some fanciers who collect their chickens in home-made boxes cover the box with a paper bag, leaving sufficient air space of course, and this does keep the draught away from the new arrivals.

Their First Accommodation

Having successfully hatched the first batch of chickens whether it be by broodies, incubators or custom hatching it is most important that suitable accommodation should be provided for the first two or three weeks. To use a parallel, chickens are similar to a new baby arriving in a household for the first time, where an extra little bit of fuss and care are made to provide the new arrival with as comfortable a home as possible. Exactly the same must be done for the chickens. Admittedly many chickens have been raised in a corner of a cabin behind an old tin sheet with no special litter or anything provided and they have thrived. It is quite possible that only the ones that have thrived are heard of— not the hundreds that have probably been lost.

Whatever happens to young chickens the main aim is to get the chickens going! There are basically three ways of rearing chickens. The first one is obviously for them to be raised under the hens that have so successfully and patiently hatched them in the nest. This is an extremely good method because the hen encourages the chickens to eat and certainly mothers and cares for them. The second is the method of raising them under a brooder. A chick brooder can be purchased in various sizes, normally some 2ft square, and the heat is provided by an oil lamp which has an adequate guard to stop any chickens catching fire. This is a good and well proved method and is certainly very useful if a lot of chickens have to be raised.

The third method is by the use of infra-red lamps. This is perhaps the most reliable way of raising chickens in that nothing can really go wrong providing the lamp is set correctly and there are no cuts in supply. At this point, we shall look at these three methods of getting the chickens 'on their feet' to see which really is the best and most suitable to suit the required purposes. But, whichever method is chosen, it must be stressed that the first home must be spotlessly clean before the chicks inhabit it. The floor must be thoroughly disinfected to kill any disease that might be there and it must be stressed that there should be no draughts. Draughts are particularly harmful to chickens and of course very dangerous if a brooder is being used because of the fire risk. There must be light and plenty of air in the building and it should be as compact and neat and tidy as possible.

Method 1   The Broody Hen

It is only human nature to let a hen rear chickens after she has so successfully hatched them but before letting her do this the first home has to be prepared. It has already been mentioned that the floor of the building must be clean and it is advisable to take an extra precaution. On the presumption that some three, four or even half a dozen broodies are using the same building, partitions will have to be made. These divisions can either be made of wood or, as can be seen in the photograph, concrete breeze blocks are useful for this purpose. A small run made of breeze blocks or wood some 4ft in length by 15in in width with a wire-netting frame on top to prevent the chicks and mother escaping is all that is necessary for the first couple of weeks. When making this run inside your building a good tip is to cover the whole of the floor with thick paper—paper bags in which food has previously been purchased are ideal. This is a second prevention against disease because the paper and all the chicken's droppings can be thrown away when the chickens leave the building, and it also provides warmth. Paper is naturally warm and is also a good insulator keeping in the warmth provided by the mother hen. The paper should be covered with some small shavings or wood chippings—preferably not sawdust because chickens do tend occasionally to eat it—and an adequate water fountain provided.

What about a small food trough for the young chickens? It is true to say that many fanciers do provide small metal troughs for their chickens and broody hens to eat from but again a much practised and successful method is simply to provide a cardboard egg tray for their food. This indeed has a double benefit because chickens like nothing better than to hear themselves pecking and this cardboard tray has the desired effect without damaging the chicks in any way. The depth in the tray which is usually filled by the eggs is now filled by food and as the broody hen is not able to scratch the food about quite as easily this does prevent wastage of a very valuable item.

Everything should now be ready for the arrival of the chickens in their first home—a good clean warm floor, a much proved food receptacle and fresh water and above all a home which is disease free. Furthermore, most important this is a home that can be extended by lengthening or widening the run as the chickens grow—the first extension to the home can be made at two weeks old and the second one at three weeks old. Then the chickens are ready for moving on to other quarters. Above all else, it is strongly suggested that the time to move the chickens into their new home is in the evening when they will tend to settle much better.

Method 2   The Brooder

Raising day-old chickens under a brooder has just one difference from raising them under a broody hen and that is that the brooder or the lamp within it provides the heat. The conditions for the chickens are similar—the house should be well ventilated, with plenty of light and it must be spotlessly clean and disease-free. The drinking and feeding utensils provided are also similar. Whilst the basic litter is the same, there should be no paper under it. And there is one most important piece of advice that should be followed if at all possible.

For generations there have been countless thousands of chickens reared under a brooder quite successfully where the brooder has been placed in a small portion of a cabin separated from the rest of the fancier's stock. Doubtless this will happen many times in the future. But there is one danger—that of fire. The fire risk can be minimised providing the brooder is controlled correctly, but if there is to be a fire surely it is better that the fire takes place in a building where no other stock are housed? If at all possible keep your brooder and the chickens under it well away and in a completely separate building from the rest of your stock.

It is vital that the lamp is adjusted with the flame giving out the right sort of temperature prior to the chickens being put in it. Obviously the chickens will find their own particular place under the brooder once they are in and providing the temperature is constant they can always get a little closer to the lamp if they are feeling a little cold or alternatively move further away from it if they are too warm. It is to establish a settled temperature that the brooder and its lamp should be lit at least 48 hours before any chickens are put under it. Again it is vital that there are no draughts about to fan the flames and no sacking or cloth near it which could eventually smoulder and cause a fire. Similarly, as previously stated, there should be no paper under the litter on the floor. This is most important and should never be forgotten.

A brooder should be placed in a confined place for the first stage of the chick's life. It is suggested that a wall, best made of concrete blocks or metal, be placed around it about 4–5in from the brooder itself which will after the first two or three days tempt the chicks to move out of the brooder for a few minutes into the open space and then allow them to go back in at their own free will. Gradually, as the chicks grow, the wall can be extended and as it is extended you can start to phase out the brooder. The best way to phase a brooder out is to turn it off for an hour the first day, two hours the second day and so on, but it should only be turned off when it is a warm and sunny day. At what age should the brooder be taken away from the chickens entirely? There is no specific answer to this because bantams need to be under it a little longer than standard-bred poultry but in the fast feathering breeds (and some do feather quicker than others) a start can certainly be made in the third to fourth week. The chickens should not be moved on to their next quarters until they have had at least a week with the brooder completely turned off.

Brooders have proved themselves invaluable over many years but they can only work if the stockman is prepared to look after them. Their oil should be checked daily and the flame should certainly be checked each evening although there is always the consolation that if the flame does go out heat will be retained for some time.

Method 3   The Infra-Red Lamp

Although this is without doubt the most expensive way of rearing chickens it is possibly the most reliable unless of course there are electricity cuts. To rear one batch of chickens it means that somewhere in the region of 550 hours of electricity will be used through one lamp and if several are being used at the same time the fancier can expect a large bill at the end of it all.

INFRA RED

However, there are probably fanciers who for personal reasons prefer to use this method. Again, the same practices of cleanliness, ventilation and light must be followed. There is, of course, still the risk of fire and for this reason again no paper should be placed under the litter.

The only other item required by fanciers using this method is a piece of metal 1ft in height which has been formed into a circle say 3ft in diameter to commence the rearing. The lamp should be hung over the centre of this enclosed space some 9in from the ground. Like the brooder, it should be switched on and the whole thing aired 24 hours before the arrival of the chickens. The correct way to use this method is gradually to raise up the lamp. At the end of the first week it should be taken up an inch and this height slowly increased as the chicks grow. In the third week it can be switched off in stages like the brooder—for an hour on the first day, two hours on the second etc. By the raising of the lamp and its switching off the chicks start to get accustomed to normal daily temperatures as they grow.

However, a fancier using this method must keep a very close eye on his chickens because if the lamp is too low there is a strong possibility that the heat could affect the chicks' feet and they will start to turn inwards. If this does happen there is a possibility that when the chicks get onto fresh land they might correct themselves but this does not always happen.

When the chicks have been reared under each of these three methods there is still a little work to be done. It is suggested that those who have used broody hens should check them for mites before passing the broodies along with their chicks on to the next stage. However, for those who have used brooders and infra-red lamps, these particular pieces of equipment are finished with until the next new batch of chicks arrives. The brooder should be thoroughly cleaned and any small points that require attention should be repaired and then stored away. Infra-red lamps should also be properly cleaned and the wiring and plugs checked to ensure that all is well for another occasion.

Feeding

It has already been mentioned that for all three methods a recommended way to feed day-old chicks is on a cardboard egg tray but what should they be fed on? It has been stressed in this chapter that the sole object with day-old chickens is to get them going, and I would recommend that for the first couple of weeks at least the faster their crops can be filled the better. Needless to say there are all sorts of foods prescribed by many fanciers who, incidentally, often have their own patent mixtures but, without doubt, one of the simplest and much proved foods for young chicks is chick crumbs. They are clean, wholesome and soon satisfy a chick's appetite. Whilst, alternatively, a good chick mash can be used, it does take the chickens, in the very early stages, longer to fill their crops and is often wasteful, particularly if being used under method (1) and the broody hen is boisterous and scratches some of the mash into the litter.

In this chapter it is presumed that the fancier rearing his own chickens for the first time will have chickens all of one colour but there is another useful piece of advice for fanciers who have perhaps hatched two or three different breeds under broodies or incubators etc. It should always be borne in mind that something can go wrong and a good tip is to mix the colours of the chickens. Rather than have all white ones under one hen and all black ones under another, each hen should have half black chicks and half white ones. Then if anything happens to one of the hens the others can be safely transferred providing of course they do not exceed the number the hen is capable of carrying. The reason for this piece of advice is that if a hen is rearing all white chicks and something goes wrong with the one rearing the black chickens then there are times when the hen to which the chicks are transferred will simply reject them. It will not allow the chicks to snuggle under her and will constantly peck them until most of them have been killed. If you visit showmen's yards in the early spring and summer you will usually see various breeds of chicks under several hens and this is the reason—in other words, it is yet another safeguard.

Once the chicks have been started off successfully under the methods described, the next thought in the fancier's mind will be when should he change his chicks' food and when should the chicks be moved? It is strongly suggested that food should not be changed until the chicks have been moved to their next quarters and then only gradually when they have settled in. The chicks should be moved at around four to five weeks old and then a growers' mash should be introduced. The best way of introducing it is to feed it damp in a morning. For the first couple of days, give separately 25 per cent damp growers' mash and 75 per cent chick crumbs. The next two or three days should be 50/50 and then 25 per cent chick crumbs and 75 per cent dampened growers' mash until the chicks are fully converted in the space of a fortnight.

The First Outside Home

Having got so far where is the next home to be? In one simple answer—outside in the fresh air and sunshine. Some fanciers rear their chicks outside on the ground from one-day-old onwards and do so very successfully. These fanciers are men of many years experience, however, and if this can be done success-fully it must be admitted that the chickens tend possibly to do a little better. However, a fancier rearing chickens for the first time wants to be certain that at the end of it all he will have managed to rear his birds successfully and it is for this reason that it is only now that the chicks are put into a movable run. But before moving them the fancier must take one or two things into consideration. He must only move his chickens outside on a warm and sunny day. When they are moved to their new homes they will undoubtedly feel strange and not only will they have to get used to the temperature outside but all the various noises of nature that attract them. Also on moving his birds from the inside quarters, the keeper must check their feet and particularly their tiny toe nails. Chickens have a habit of picking up wet litter on their nails which eventually dries and becomes what is known as 'muck buttons' and these must always be removed. It is also suggested that part of the run in which the chickens are to be placed should have a cover on so that if they are outside in the run and a heavy shower comes then they have something to get under and out of the rain. Some fanciers use a large piece of glass as their cover because this not only gives a shelter in case of a shower but also provides warmth when the sun's rays shine through it.

The Hay Box

What sort of accommodation is required for the chickens as their first outside home? Firstly, of course it should have a good sleeping compartment with a strong wire floor which should be securely fastened to prevent the chickens being taken by vermin and it should also have a small run in which they can move freely around during the day. This unit is often called a 'hay box' as seen in the photograph. It has a separate door which is formed as part of the roof, and a sliding pop hole for the chicks to move in and out of into the run. The run should also have two separate doors should any of the chicks have to be caught. These hay boxes have proved to be very useful over many years and can generally be purchased at reasonable prices.

With a hay box the fancier should ensure that the wire netting covering is strong and vermin proof and another useful idea is to put barbed wire all around the bottom of it. Whilst this will not prevent rats or badgers burrowing under, it will, nevertheless, put them off, particularly if they happen to scratch and cut their nose on one of the barbed wire spikes. Again this is a method that has been proved with great success. The fancier should also ensure that the whole thing is well creosoted and it is vital that this is done long before the chickens are put in it. A good idea is to creosote all the hay boxes and equipment such as this which will be required for rearing chicks in the very early spring so that it is dry and free from all odours before being used. But most essential in the small house in which the chickens spend the hours of darkness is the good thick wire mesh bottom to prevent the chicks being taken by vermin and which the droppings can pass through to avoid disease.

Besides looking after his stock as far as feeding and watering is concerned, it is probably now clear that the fancier must always have in his mind the possibility of vermin being present when he puts his chicks outside. It is because of this that it is recommended that hay boxes or similar pieces of equipment should be purchased from a local poultry house builder rather than a fancier make his own. Agreed, many chickens have been successfully reared in home-made runs but these do have the disadvantage of not being quite as vermin-proof as those which are specially made for this purpose. Generally, home-made chick runs consist of a small coop with a separate movable lean-to run instead of being a compact unit like the hay box. These home-made runs are ideal for very young chickens but can prove to be the wrong piece of equipment as the chickens grow.

For the first 48 hours a close watch should be kept on the chickens in their new home and though the broody mother will scratch for them and generally care for them, the first problems usually occur at dusk. The broody, used to a hay box will go through the little pop hole into the sleeping compartment quite easily and either take a few chickens with her or none at all. Alternatively, broodies often snuggle down in a corner of the run with the chicks under them and this too is dangerous should a heavy thunderstorm take place during the night. Therefore, it is important that the broody (if, of course, one is being used) and the chickens are seen safely into their sleeping quarters. It is then just a question of sliding the hatch closed. After a few nights they will get used to going in on their own.

As the days go by the run will become devoid of grass and in order to stop the ground smelling and going sour the whole run must be moved at least every four or five days and possibly a little bit oftener if the weather is wet. Continuous heavy rain does tend to make the run a little swampy especially when the chickens are growing rapidly.

After the chicks have been in this type of accommodation for two or three weeks they will be seen to be growing and feathering rapidly and certainly in most breeds cockerels and pullets can be identified. You will notice that they take less and less interest in the broody hen and it is at this stage that the broody can be withdrawn and put back in her normal run. Before putting it back, however, dust to prevent lice and mites and also check legs to ensure there is no scaly leg—this is not a disease but a small mite which lives under the scales of a hen's leg. A very good remedy for removing this is sulphur ointment which can be purchased from any chemist and a good time to rub it on her legs is before returning her to her pen. Doing it then will ensure that when the broody is used again for the rearing of chicks her legs will be clean and she will not pass on these little mites to her new flock.

Where the chickens have been reared under an infra-red lamp or brooder the chickens must have been completely off heat for at least a couple of weeks before putting them outside. They must never be put out during a cold spell because this will only ask for trouble. Possibly birds reared under these conditions need a little more attention in the first few days than the chicks that have their mother broody still with them, but apart from this there is no difference whatsoever in their rearing and management.

Usually, young stock can remain in hay boxes until they are in the region of nine or ten weeks old and bantam chickens certainly longer if so desired. By the time the stock reach this age they will be growing rapidly and if boredom is allowed to take place, particularly through not moving the runs onto new ground, then trouble such as feather-pecking will break out.

The Night Ark

It is at nine or ten weeks that the chickens should be moved to their third and final rearing premises. The chickens, now that they are fully able to look after and fend for themselves, should then be housed in a night ark. This is best described as a miniature poultry house with slatted floors through which the droppings go. The slats can be removed for cleaning when the chicks vacate the house. At this stage the birds are no longer chickens but growing cockerels and pullets and they normally spend the rest of their growing days in a night ark until at the age of 20 weeks, when the pullet becomes a point of lay bird, they can be moved into their main quarters. Another reason why it is recommended that they be housed in a night ark with a slatted floor is that young birds at this age tend to perch and if the perches are not the right shape the birds will develop crooked breast bones. If you intend to show this stock then you will be

Plate 10 The author, in his trophy room, with some of his many cups and prizes

Plate 11 The Penning Room: note the clean litter in the pens and the individual water troughs, the tidiness of the room with a clean swept floor and brush, shovel and bucket neatly placed. There is plenty of natural light and an electric socket

gravely disappointed because crooked breast bones are classed as a serious deformity in exhibition birds.

The feeding of the birds in the night ark continues virtually as before. Growers' mash, nicely dampened, should be fed every morning and in the evening a few handfuls of mixed corn thrown around the area of the ark so that the birds have got to scratch for it to find it. All this of course helps to keep them out of trouble. Another point that should be borne in mind is that as these birds grow they will need extra feeding troughs and cramming (or crowding) should not be allowed around the troughs—it is far better to have one trough too many than one trough too few. The water fountains that have been used for them as chicks and during the time they spent in hay boxes will now be found to be too small and either a couple of extra ones or one of a far greater size should be used.

Some fanciers, because of limited space use the same small building to rear their stock from day olds to point of lay birds. This is quite acceptable and is a method that has been practised with much success but there are certainly disadvantages to it. Firstly, it means that only one lot of chickens can be reared at a time in this building. But more important is the question of the ground surrounding the building that gives cause for concern. After having birds running on it for the best part of 20 weeks it is bound to have become stale and sour and a disease trap. Coccidiosis, one of the diseases that every stockman is concerned about and mentioned earlier, often develops and can be seen in birds reared in this manner. Not only will one batch of birds suffer but the ground would certainly not be as sweet as it should be for when the following year's chickens come along.

Laying Pullets

It is at the age of about 20 weeks that the fancier really starts to sort his birds out. If he is rearing birds for egg-laying the next move is simpler than for the fancier rearing birds for show purposes. Always bearing in mind that the successful stockman is planning ahead with the accommodation for his stock the move of point of lay pullets to their final home is relatively simple. Many fanciers simply collect all their point of lay pullets in a crate at night from their respective night arks and put them on the perches of their new home. Fine, but why not take an added precaution? Time and again in this book the question of mites has been mentioned. Birds will only lay as well as you look after them, so why not take an extra little bit of time and trouble to ensure their comfort? Dust or spray them to prevent lice and fleas as you put each one into its new home and, better still, put a little nicotine sulphate on their legs, as described on page 42.

Potential Show Birds

Problems for the showman are somewhat different and it is at this age that he can really tell whether his birds are going to be of any use or otherwise. Many showmen have a good idea at a much earlier age although this does depend very much on experience and particularly the breed of the birds. Birds that have detailed markings have really got to be reared to this age before establishing their true value. So now is the time that the showman has to sort out and decide which cockerels and pullets he wants to keep. Generally, the pullets are moved on in the same way as the laying birds and the cockerels are kept away from them. By doing this the feathers will not become damaged by over-exuberant cockerels and the birds will be kept in the best of condition for show purposes.

The problem with young cockerels is that they tend to fight. A good practice to adopt is to select the ones you want to keep, put a ring on them and put them back in the same pen within minutes. If a bird is kept out for even a day and then put back, fighting will certainly start and the potential show bird or birds damaged. The showman should make every effort to get rid of his spare birds quickly, otherwise they tend to become a nuisance and fighting will ultimately start. If cockerels are left together in the same accommodation in which they have been reared, fighting will certainly be delayed for some time. As the showman sells the spare ones, one by one, he will eventually finish up trouble free and with just the ones left that he wishes to keep for himself. It is at this point that he can then put them into other small runs or even in training pens to settle down.

Eggs
The final point to consider is the feeding of laying birds and the conversion from growers' food to layers' daily food. The change-over is exactly the same as changing from the normal chick food to growers' food described earlier—in other words, do it gradually over a period of some fourteen days and start with seven parts of growers' mash to one of layers' mash and then six of growers' mash to two of layers' mash etc. The change will not be noticed and should be superbly timed with the pullets coming right onto lay. Keep an eye open during the first few weeks for soft-shelled eggs. Generally these are caused by over-production but a few handfuls of grit are always a help. Above all at this time it is vital that the pullets are kept relatively quiet and all forms of pets should be kept well away from them until they have settled in. If strangers come to see the birds then they should be taken quietly and slowly into the pen because if the point of lay pullets are startled it will have the effect of knocking them off lay right away. Of course for the first few weeks, eggs will be somewhat small but gradually they will increase to normal size after several weeks production.

There is certainly no hard and fast rule as to how many eggs point of lay pullets should lay in their first year—this is a matter for the experts namely the commercial breeders. Anyone who has decided to rear poultry purely for the commercial purpose of keeping the home supplied with eggs is well advised to buy his stock from breeders whose birds have a fully established laying record.

6

DUCKS, GEESE AND TURKEYS

No poultry book would be complete without a chapter, no matter how short, on ducks, geese and turkeys and whilst this book is written basically for beginners nevertheless mention of them must be made. There are many advantages and disadvantages to keeping any of the above types of poultry and before even considering keeping them a fancier should have already built up his basic knowledge of 'cocks and hens'. They do indeed differ from ordinary laying hens or indeed exhibition stock for that matter and a thorough examination of the conditions and area available must be made.

Ducks

Ducks are generally to be found in six breeds, namely: Aylesbury, Khaki Campbell, Indian Runner, Rouen, Buff Orpington, and Muscovy. They all have their particular qualities which are listed as follows:

Aylesbury A useful layer but more regarded as a table duck and ducklings at nine weeks of age have been known to weigh 6lb weight.
Khaki Campbell An excellent layer and possibly the greatest forager of all ducks.
Indian Runner Another good layer.
Rouen Normally regarded as the showman's duck. A typical water fowl in every sense of the word. On the other hand a good table duck but the flesh is much darker.

Buff Orpington A good all round duck but possibly one of the least popular.

Muscovy Bred specially for the table.

Anyone considering keeping water fowl must have soft land available and plenty of good growing weeds. Dry land is not beneficial as far as ducks are concerned.

Adult Stock

If it is intended to keep ducks commercially, for a living, a good supply of water must be provided, particularly natural water, such as a pond, and if possible the running water of a stream. Many ducks actually mate in water and consequently if they are kept penned up on dry land fertility will be low. Good dry sleeping quarters are most necessary and they should also be well ventilated. Ducks do not perch and consequently the floor must be clean and free from dampness and their sleeping quarters be some distance away from the stream or pond. It is for this reason that adult ducks are fed only at night—in the early evening they should be tempted from their pond to their sleeping quarters by the rattle of a bucket containing good sloppy mash, and they will usually make their way in one long line to their food troughs. A wire netting run of no more than 18in high should surround the sleeping quarters to prevent the ducks getting back to the water after feeding. As for chickens, sleeping quarters must be secure because of the danger of foxes and other vermin.

Ducks always lay early in the morning and it is for this reason that they should be shut in at night. Failure to do this will result in eggs being laid in the middle of a stream and washed away which, of course, is simply no use to anyone.

Whether young ducks or adult stock are kept it must be noted that sun is something they do not like so shade must be provided even if it is just some corrugated sheets or boards which they can go behind to get shelter. Water must always be available be it only a large trough in the case of growing ducklings so that they can regularly dip their heads in. This practice is called dibbling. Young ducklings do suffer with their eyes getting sealed up and by dibbling under the water it helps to clean their heads and keep their eyes free.

Young Ducklings

The incubation of duck eggs is 28 days and normally a broody hen can rear in the region of six to eight ducklings. A broody could, in the very early stages, rear more but the ducklings' growth rate is so tremendous that at a fortnight old they may no longer need heat from any source to keep them warm, and the broody is quite incapable by then of caring for any more than a maximum of eight. Ducklings have a very fast weight gain, particularly if they are fed correctly and the old phrase, 'little and often' should be carried out. Soft mash is best for them and as they grow older mixed corn can also be fed to them but it should be thrown into the water where dibbling takes place. Whilst young ducklings can certainly be reared to nine or ten weeks of age without swimming water, stock from which it is intended to breed the following year should have access to a pond or running water after this age, in order to build up their stamina and bone structure for breeding.

There are certain advantages in keeping waterfowl providing the conditions are right for them. For instance they do not scratch; they clear a field, orchard or garden of all the slugs and in winter-time, they are relatively easy to keep; they take very readily to intensive conditions and this is a great advantage particularly if it is ducklings to nine or ten weeks of age that are being reared; they are certainly much quieter than poultry and could well be considered for any small unit; as far as the showman is concerned they have the great advantage of being a species of fowl which is long lasting and many waterfowl exhibitors have won championships repeatedly through selecting the right bird or two which have kept them in the forefront for many years. This of course has the benefit, particularly as far as finance is concerned, that there is no need to rear many replacements.

Geese

Unless there is ample ground available geese should probably not be kept. Geese are great grazers and are certainly of no use whatsoever to the small poultry keeper who has only small runs and a very limited space available.

If accommodation abounds geese are certainly useful. They will keep orchards well mown and need very little attention. They are easily managed and not affected by adverse weather conditions. The only housing required is somewhere to provide them with shelter—three walls and a roof are all that is required and provided it is situated in an easily accessible position the geese will be perfectly happy. However, it must be secure at night.

When breeding, it is normal to mate two geese with one gander and the resultant eggs will take 28 days to incubate. Four eggs can be placed under a broody and when hatched can be put outside in the small run immediately. The goslings should be kept penned up in their run for at least two or three weeks, moving the run each day. After this, the goslings along with the broody should be let free to graze at their own will.

Geese are, of course, known to sit their own eggs. Generally speaking the goose will lay anywhere from 15 to 20 eggs before showing signs of wanting to go broody. A good way to tempt a goose to sit is to leave the eggs in the nest and this will usually result in her covering them, after which she can be left to rear her brood in a straightforward, protective manner once they have hatched. If you have the space, geese are certainly birds that need little attention and are virtually trouble free.

Turkeys

Providing the conditions are right there is certainly no harm in the poultry keeper considering the rearing of ducks and geese but what about turkeys? This book, dealing with poultry keeping for beginners, is really not the place for discussing the advantages and disadvantages of keeping turkeys. Sufficient to say, therefore, that until plenty of experience is gained, turkeys should not be considered. Turkeys still have some of their native instinct left, as proved by their willingness to perch high up in trees when left free to roam. Young turkeys too are not easy to rear and are susceptible to a disease called blackhead which turkey breeders, never mind poultry fanciers keeping them for the first time absolutely dread.

The land too should be right for turkeys and, if they are allowed to roam, dry land is an essential. They require special housing if they are being kept commercially and as such a complete and detailed study should be made of this species of fowl long before considering keeping them, and specialist books are available for this.

It must be hastily added, however, that the poultry keeper who is fully acquainted with the correct way of rearing and handling turkeys can find it very rewarding particularly as far as finance is concerned. But as with the other two species experience should be gained with other types of stock before turkeys are considered.

7

THE COMPETITIVE SPIRIT —
EXHIBITING AND JUDGING

As time goes by, the poultry keeper who only a short time ago commenced keeping poultry develops into an 'apprentice stockman' and undoubtedly his involvement with his stock will be growing too. Instead of making a routine visit to his local summer show he will make a special note of the date in his mind. Instead of wandering aimlessly around the showground, his first priority on arriving will be to visit the poultry tent. He will be keenly interested in the exhibits and anxious to broaden his outlook and knowledge. He may well think in his own mind, without any real knowledge of exhibiting, that he could have won a prize himself! Alternatively, he may even feel that his own eggs are browner than those that have won.

Well, could his birds win and his eggs gain a place? Doubtful!

In this chapter, devoted to fanciers who wish to commence showing poultry, the basic rules and routines will be discussed not only for showing birds but also for showing eggs and, to a lesser extent, table poultry. I do not propose to go into the various standards of individual breeds because they are all so different and specialist books are available on the subject. However, a fancier who decides to embark on a showing career must remember at the outset to walk before he can run. Like everything else, an apprenticeship must be served. A golden rule is that a fancier should only exhibit birds or eggs that are a credit to him and a credit to his stock. Above all, whatever the decision of the judge, a fancier must learn to accept it. Indeed, when he completes his entry form he must sign to abide by the rules and regulations of the show society and accept that the judge's decision is final even though there are times when he possibly considers himself unlucky not to win. On the other hand, there are times when the exhibitor and indeed others will consider himself lucky to win. It has been the misfortune of many fanciers at the start of their showing career to be very lucky with birds they have bought so that when they have come to exhibit birds bred by themselves, which are possibly not as good, their enthusiasm has been dampened and eventually they have disappeared. The best advice is to go slowly but surely and climb the rungs of the long ladder one by one. If you do this, great satisfaction will be achieved particularly when you show good stock you have bred yourself.

Members' Shows

How does one go about showing? Usually anyone interested in showing should join their local poultry society. Attending the monthly meetings that most of them organise he will be told or see notices that a members' show will be held two or three times a year, possibly culminating in one large open show which usually takes place during the winter months. A fancier should start in this way, exhibiting at his local poultry society's members' shows rather than jump into the deep end at one of the larger shows where he will be exhibiting his stock in competition with fanciers who, in many cases, have been showing birds for years.

Judges

Without doubt, virtually every judge will help if asked politely and decently why he has placed your exhibits where he has done, and particularly if he can readily see that you are a new fancier and anxious to learn. After all, a beginner of today is the fancier of tomorrow. Very often judges will help beginners not only by pointing out the faults of their birds but also taking the first and second birds that he has placed and pointing out their good points and the reasons why the first one was preferred to the second. He could well go further and give advice on breeding and whether the particular stock the fancier has exhibited is worth breeding from. It must be appreciated that the knowledge given is based on many years experience of breeding and showing and is priceless—the new beginner should note it in detail. Obviously not all judges think alike in detail but, generally speaking, the opinions are largely the same as far as basics go.

Breed Clubs

After competing at one or two members' shows the new exhibitor will no doubt have found the name and address of his breed club secretary, and will have been in touch with him to enrol himself as a member of the club. The breed club secretary, an experienced showman, will go out of his way to help him and will send him details of the laid-down standards of the particular breed. These standards are, incidentally, all formulated by the breed club in conjunction with the Poultry Club which passes the standards and any alterations to them at its council meetings. The Poultry Club, of course, is the governing body of the exhibiting of poultry just as the Kennel Club is the governing body of the dog world. On receiving the standard, the new fancier will interpret it in relation to his own birds, and if he couples this with the advice from the show judges he should start to formulate the right sort of opinions as regards his birds.

Rules and Routines

Perhaps the time is opportune to discuss aspects of showing from the very first schedule that a fancier receives to the time when he will eventually secure his first prize and of all the prizes that he will win in his show career the first one is indeed the most treasured.

Schedules

Receiving a schedule, an exhibitor will find that the classes are generally split up into exhibition large fowl, utility large fowl, hard feather bantams, soft feather bantams, possibly waterfowl and eggs. He will note that the classes read either male or female which means that birds of all ages providing that they are the right sex can be exhibited in these classes. However, he must take careful note of whether it says cockerel or pullet only birds, which means that only birds hatched after 1st November in the current year of breeding can enter these particular classes. Again he must make a careful note of the various colours of particular breeds for it is important that the birds are entered in the correct coloured class. Note should undoubtedly be made of the prize money and entry fee. It is important for some shows that the competition lists on the entry form breed clubs of which he is a member in case of 'special' prizes being allocated. Particular note must be made of the date the entries close—no secretary likes receiving late entries and many simply refuse them. It is only fair to the show secretary who has a great deal of work to do in organising the show to let him have entries in good time.

Looking at the classes in the schedule, the large fowl classes speak for themselves because these are for standard-bred birds and there the various breeds will be listed. There is also a possibility that a class for cross-bred hens may be included. This is particularly relevant to shows in the north of England where such classes have become somewhat of a tradition and include birds of many shapes, sizes and colours. Needless to say, it is not necessary to produce any proved egg-laying records because the birds are judged as the likeliest layer. It may be noted too that there are classes for utility poultry and these are for birds which are pure bred without being exhibition bred, and are capable of producing good table birds or, more important, birds which are typical of the breed and will be consistent egg producers.

A new exhibitor is often confused by the terms hard feather and soft feather. Basically what this means is that all game bantams are hard feathered and the remaining breeds are soft feathered. Indeed the classes for bantams are generally far more numerous than those for large fowl but again the intending exhibitor must be extremely careful that he enters his birds in the correct classes and fills in his entry form correctly—if he decides to enter two birds in a class, two lines must be taken up on the entry form.

At the bottom of the schedule are usually listed the egg and table poultry classes and, here too, careful note must be made of the number of eggs required per plate. Usually, plates are provided by the society staging the show but it is always useful if you are exhibiting eggs to take along one or two spare paper plates just in case none happen to be on the tables on arrival.

The final thing that all fanciers must note is the time when the exhibits can be taken away from the show. There is no such thing as simply having your birds judged and then putting them in their respective hampers and away home. It is agreed that a show is staged primarily for fanciers to enjoy the competitive spirit but it must be remembered that the public want to see the birds as well. Indeed, not long ago the new beginner was himself just one of the general public so keen to see the exhibits staged. On no account, therefore, must any birds be removed from the show before the stated time.

Having decided what to enter, the fancier will then complete his entry form and post it off to the show secretary. As each day passes and as the show comes closer he will become keener but it must be remembered that the shows are not won simply by taking the bird off the perch the night before, putting it in a basket and then into a pen in the exhibition hall. Many fanciers start this way but they very soon find out that is is not the way to win. Birds intended for show must be prepared properly. After having established that the birds to be shown are the correct type, shape and have the right markings they must appear to be fit and healthy—and then comes the real work. They must be carefully prepared by washing but all intending exhibitors should realise that washing will not make a poor bird a good one . . . however, it could well make a good one into a champion.

Washing

There is an old saying amongst fanciers that 'first prizes are won at home' and this has been proved time and again. First of all the birds must be in condition and secondly washing plays a very important part. When you start to wash birds never do it in a hurry—it is often a long tedious job which requires plenty of care.

Not every breed requires washing before it is shown. Generally speaking, it is the soft-feather varieties, particularly birds that have white in their colourings that need it. Washing birds is something of an art and the first time the fancier attempts it he will wonder to himself whatever he has started. The bird will probably flutter and splash water all over the place and the fancier will be so alarmed by it all that he will be terrified of drowning the poor thing. All fanciers wash birds basically the same way although some may handle them a little bit differently in detail. Some, too, dry their birds in different ways than those described in this book but if the simple routine ways are followed for a start then a fancier can adjust them as he progresses.

Washing a bird for a show requires three lots of water. The sink or other wash tub should be filled to a depth of some 6–8in with warm water and the bird placed in it for half a minute in order for the water to cover all the feathers—needless to say the head must be kept well above the water mark. After the water has reached all parts of the body and the feathers are well soaked washing can commence. First of all the wings must be given a thorough wiping with a bar of soap or with a good lather on the hands. Next comes the tail and afterwards the body with a final wash of the hackle and breast. For most of this time the bird's legs will have been in the water and will be thoroughly soaked. At this stage a fine nail brush should be taken, the legs given a good scrubbing and all the dirt removed from under the scales— this is particularly applicable should the bird be two years old or over. Finally comes the head. With a soft soapy hand the comb should be carefully washed and then the lobes (if it is a breed that carries them) and then the wattles and face. You cannot really help getting soap in the bird's eyes but, of course, the less that manages to find its way there the better.

At this point the fancier will find himself holding a very wet and bedraggled soapy bird looking completely lost and giving the impression that it will not live to see another day! The fancier will particularly notice the filthy water that has resulted from the washing and he must get rid of this and replace it with the same amount again of warm water. The soap should then be completely rinsed from the bird and in order to ensure that the washing has been a success, this second batch of water should then be thrown away and replaced with the third and final rinse. Again the water should be warm and the bird thoroughly soaked and any soap left should be squeezed from the bird's feathers and on this rinse the head, too, should be given a gentle sponging of warm water. It cannot be emphasised how important it is to keep the bird's head away from the water at all times except when it is actually being washed.

Drying

Many fanciers have their own particular methods of drying. Some use hair-dryers very successfully but again this is a long arduous task especially when up to six or ten birds have to be washed the same evening. In fact it often develops into an all-night sitting! Other fanciers have specially made boxes with a wire netting front in which they place their birds and then position them in front of a good fire where they leave the birds to dry. As others are washed the boxes are moved to one side and replaced with another one until the job is completed. These boxes can be left in front of the fire all night and the following morning the birds will be virtually dry.

Many fanciers who have a good knowledge of carpentry and engineering have built 'drying machines' to their own specifications. Generally, these take the form of a unit which blows hot air into the bottom of some kind of box, and then rises through spars made of metal on which the bird is standing, and blows the feathers dry. This type of machine has been proved to be particularly successful for two reasons. First of all the bird can move round and spread its wings at its own free will and preen itself, and so does tend to dry very evenly. There is also the added advantage that any droppings go through the metal spars and onto the floor of the cabinet. If care is taken, more than one bird can be dried at once in this form of equipment particularly if suitable divisions are inserted. The only disadvantage in using something of this style is the need to check the consistency of the temperature of the hot air being blown in, because if care is not taken it could have a burning effect.

All fanciers have their own particular ways of washing birds formulated over several years of experience, and so too are some of the additives placed in their birds' wash. Some breeds really are shown 'whiter than white' and this is often accomplished by the addition of some particular item which is a closely guarded secret of the fancier. This, of course, can only be discovered by trial and error and if some mixture is recommended it should never be tried out for the first time on birds which really matter. The fancier should, of course, ensure that whatever he is recommended to use will have no harmful effects on his stock and particularly on the birds' skin and eyes.

After the bird has been successfully dried, it should then be placed in a clean pen in a specially built penning room. An adequate supply of clean shavings or sawdust should be ready for the bird's arrival in the pen and so too should clean drinking water. Birds that are intended for showing should be continually looked at whilst in their training pens and regularly cleaned out. It is absolutely no use washing a bird for a show and making a splendid job of it if it is then left in a dirty pen. The old phrase 'it is no use putting new wine into old bottles' certainly applies as far as showing poultry is concerned.

The Show Pen

It is easy to tell birds that have never been shown before but every bird has to start somewhere. Far too often birds are seen at shows fluttering about the pens where they are completely unsettled and ill at ease. How then does a fancier get his birds to settle down and show themselves as true specimens of their breed when at a show? The answer is quite simple. He erects a small cabin or divides part of a larger one and in it erects anything from six to a dozen show pens exactly the same type as used at the shows. Some two or three weeks before the show, he places the birds that he intends to exhibit in them and after three or four days they get used to these strange surroundings and start to settle down. As the fancier opens the door to feed the birds they also get used to his hand going into the pen. He should then try to tempt them to feed out of his hand so that they do not get alarmed when the door is opened. The best time to do this is at dusk when the birds can be quietly handled and moved around the pen and they will soon start to settle.

Birds that have been shown several times soon get used to being put in one of these exhibition pens and indeed the fancier will find that as time goes on they simply stand there and wait to be picked up. There is nothing worse than having a trained bird in a show pen which when the judge comes along is startled, and flies all over the place covering him with sawdust. It is only human nature that this kind of thing will be frowned upon.

These pens are also most useful after the birds have been washed. Clean birds can be put into a clean show pen with new litter and remain there until the morning of the show. During this time, of course, great care must be taken of them and they should be cleaned out frequently.

Hampers
Shortly the great day will arrive—the day of the show. Again, cleanliness is the operative word. Birds arrive at shows in all sorts of containers and sometimes in all sorts of conditions, but how wrong it is to send a good clean bird that has been superbly prepared in some tatty old box with stale litter in the bottom. Whether one is using a recognised wicker-work show hamper (Figure 10) or cardboard cartons with plenty of air and ventilation there are two things that must be observed. Firstly, the container should be large enough because if birds are cramped into a small space especially on a long journey they will get warm and it has been known for some to suffocate. The other essential is, of

course, that the containers must have good clean, fresh litter in the bottom.

SHOW HAMPER

The word cleanliness also applies on arrival at the show. The fancier will know which pens have been allocated to him on the sheet which he will receive from the show secretary, and a sound piece of advice is to wipe the inside of these pens before the birds are put in. The pens at many of the country's shows are stored away in all sorts of places between shows and though they are all disinfected afterwards they do gather dust whilst they are stored and a failure to wipe down the inside of the pens has resulted in many a white bird showing a cast of black before it is even judged. Care too should be taken that there is an adequate supply of litter in the bottom of the pen and especially that the pen is secure.

Therefore, with a bird in good condition, that has been washed and dried correctly to show off the natural bloom it possesses, that has been brought to the show in a container of suitable size and placed in a clean pen with clean litter in the bottom all that is required now is the examination by the judge. Compare this routine with that of those fanciers who, the night before, pick a bird off a perch, put it into a crate, bring it to the show and simply throw it in the pen and one will immediately realise who stands the better chance of winning. Some fanciers never learn but those who are prepared to go through this task year in, year out are the ones who climb up the ladder and eventually to the top. Surely if a show is worth going to it is worth showing any exhibit in the correct and proper manner.

Prizes
After the judging of the classes has been completed the judges will award the best of breed rosettes which are normally donated by the breed clubs and societies. The award of these special prizes varies from breed club to breed club and also whether the particular judge is a recognised club judge but, in the majority of cases, they are only awarded to the members of the respective breed clubs, hence it is certainly an advantage to be a member. After the breed special prizes have been awarded, the best hard-feather and best soft-feather bantams are chosen and so too is the best exhibit in the large poultry and waterfowl section. And then comes the supreme championship. This is the choice of the bird from the above four categories and the judges consider the merits of each one before placing their verdict. Should the judges opinion be equally divided, an independent judge is usually brought in to make the decision. To win at a show can be very rewarding from one's own personal point of view, especially if the bird has been one of the current year's breeding. It proves that the fancier is on the right lines and also gives him great encouragement for the future because whilst not all pullets and cockerels make good second-year birds many of them certainly do. The fancier will certainly feel great satisfaction, especially if,

little by little an array of rosettes, prize cards and even silver trophies are gathered. As time goes on a judging engagement may be offered and subsequently thought be given to taking the Poultry Club judge's examination.

But all this happens at the show—what should happen when the birds return home? Firstly and foremost, as the birds are taken from their containers they should be given a small dusting or spray with anti-mite or lice powder. Not all the birds at the show will be free of these pests and they are often passed on from one bird to another, especially on a warm day. If this

PEST SPRAYING

dusting is not given and birds have picked up some of these insects, before long, an entire cabin or penning room could well become infested. The birds should then be put back in their show pens if they are to be shown again or, alternatively, back into their breeding pens or runs. The containers should be stacked away in a good dry shed and not simply thrown to one side for other birds to perch on. It is probably hard to believe that the condition of some of the hampers brought to shows is absolutely appalling—they deserve to go on a bonfire rather than in an exhibition hall. Many of them are never cleaned from one show season to another and little wonder that if a fancier cares for his hampers in this way this is reflected in the care and preparation of his birds and, consequently, few prizes find their way to his yards.

What Makes a Winning Plate of Eggs?
Even if the fancier does not keep exhibition stock there is no reason why he cannot take an active part in his local show or poultry society. Indeed there is a very special place for him should he wish to have an interest and competc—the egg classes. Most shows throughout the country include in their poultry schedule classes for hen eggs, bantam eggs and occasionally duck eggs and most of them also include a class for the contents of an egg. Usually the eggs are staged on cardboard plates or separate clumps of sawdust and the following is a general classification:

Plate of three white hen eggs
Plate of three brown hen eggs
Plate of three tinted hen eggs
Plate of three white bantam eggs
Plate of three any other colour bantam eggs
Contents class

Occasionally there is a class for the ideal egg, hen or bantam, and also an occasional class for duck eggs in an area where there are several waterfowl enthusiasts.

What does a judge look for in a plate of prize winning eggs? First of all, the exhibitor must provide eggs that are of a good size. They must also have a good shell, no rough spots on the broader end of the egg, and no wrinkles. Needless to say, they must all be of the same colour and the whites must be white (not cream) and the browns must be brown. The eggs must have a

STANDARD

Distinct narrow and broad ends, smooth even shell texture free from porosity.

ROUND

Too large air space indicating staleness.

OVOID

No distinct narrow and broad ends. Misplaced air space.

NARROW

Straight sided.

RIDGED

Usually double yolked.

STANDARD & FAULTY EGG SHAPES

good texture to their shell, not be highly polished, and should be shown as fresh as possible so that the natural bloom appears. They must all be of the same shape and this is where care must be taken. Eggs that are long and narrow are of no use, neither are dumpy eggs. The ideal type of hen egg is broad at the top coming down to a narrower, nicely curved point at the bottom. An ideal show egg should certainly not be pointed at the narrow end. The judge will take very great care and examine each egg and he will most certainly look first to see that they are all the same shape and all identical. In order to achieve this it is, of course, beneficial to collect the eggs from the same hen and if only 12 or 15 are kept in a run this can easily be done. A fancier interested in showing eggs generally keeps the type of breed which will produce him the colour of egg that he wants—Minorcas and White Leghorns are kept to produce white eggs and such breeds as Marans and Welsummers are kept for the brown eggs. Tinted eggs are produced by a great number of breeds and, indeed, crosses and these are of course more prevalent.

Fanciers showing eggs will always come across visitors to the show who appear to be absolutely amazed at the colour of some of the brown eggs that are exhibited and frequently one hears remarks that these eggs must have been dyed or specially prepared. Indeed this is not the case because the breeds which lay these wonderfully coloured chocolate eggs are not the most prolific layers and, consequently, are not kept on any large commercial scale, hence their very rare appearance in the shops. In the section for plates of three (or four) of the differing colours the judge rarely breaks an egg and he usually leaves the cracking process until the contents class.

In the contents class judgement is entirely on the contents of the egg. A good contents egg should have a rich-coloured yoke that stands well up from the white (the albumen) and three visible layers of white around the yoke. Two little pieces of what might be termed gristle which keep the yoke suspended in the egg should also be clearly visible. These are called the chalazae. Above all, there should be no blood spots in the contents of a cracked egg. Often, of course, the judge will see watery whites and pale yokes and there is no doubt that the contents of an egg can certainly be influenced by feeding. Birds which are on perfectly natural pastures picking up all things that nature provides for them are the ones which lay eggs with ideal contents for the show bench in comparison with birds kept indoors for the greater part of their lives and receiving most of their food in artificial form.

Exactly the same rules, regulations and standards apply to the exhibiting of bantam eggs, except with respect to size. Bantam eggs must not exceed 1½oz. As with the larger birds, a fancier who keeps ordinary pure-bred or even cross-bred bantams is able to compete. Just because a bird is an exhibition bird, it does not necessarily mean that it produces good show eggs—in fact, just the opposite, because some of the best eggs exhibited are from ordinary pure or cross-bred birds.

The fresher the eggs are, the more chance they have of winning —eggs that have been laid for some time appear to be dull in their shell and indeed the judge can always identify them by a method known as candling. In the broader end of every egg is an air space and the older the egg, the larger the space becomes, so consequently a judge with a strong light shining through this space can determine the freshness of the egg.

The art and interest of egg showing is that it enables every fancier throughout the country to take some kind of active part in his local show, or indeed shows further afield, irrespective of the type of birds that he keeps.

Table Poultry

Years ago, most shows used to include a class or classes for table poultry along with their egg classes but sadly these are becoming fewer and fewer in number. Briefly, the exhibiting of table poultry prepared ready for the oven is based on common sense. Needless to say, all the birds should be killed, plucked, dressed and trussed and in many instances are shown with their legs and feet still on. Common sense will tell exhibitors and judges that birds with crooked breast bones or with flesh torn when being plucked will lose points. So too will birds that have any type of growths or cysts on them. What the judges are really looking for is a bird that he would enjoy seeing on his own dinner table! A bird that has a good quality and a good quantity of meat with a good long breast bone covered with plenty of flesh and a broad and flat back to go with it, in addition, of course, to two jolly good drum sticks either side.

Providing a bird fulfils these points, is properly presented and is of an appetising colour then it has every chance of winning its class. Presentation of table poultry plays a big part in gaining prizes, and the liver, gizzard and neck should also be suitably shown. Without going to too much fuss the use of the odd sprig of parsley can help to set off the whole exhibit.

It is sad to see these classes going but in view of the high standard now demanded, and correctly so, by the health and hygiene authorities it can be appreciated why they have generally disappeared. Surely there is nothing worse than to see a row of good cock chickens at a local summer show in a hot tent with hundreds of bluebottles buzzing all round them. Hardly a sight to cherish and certainly not an advertisement for the poultry fancier or indeed the poultry industry. If table poultry have to be shown these days then suitable hygiene safeguards must always be provided.

Your Local Poultry Society
Earlier in this chapter one of the benefits of joining your local poultry society was highlighted—that of becoming familiar with the show routine, the correct way to enter shows and how first to commence exhibiting. But your local poultry society is generally not organised solely for the exhibitor. It is organised for everyone who keeps poultry whether they are admirers of exhibition stock, commercial stock, table poultry or indeed water-fowl—there is a place for all of them. Agreed, most poultry societies hold their large annual shows, supported, probably, by a couple of members' shows throughout the year but there are many other activities as well. For instance, film shows and talks on a wide range of topics are given at the regular monthly meetings during the winter. Very often the representatives of feed suppliers are only too happy to come along and give a talk on the advantages of using their particular foods, perhaps supported by slides and occasionally a film. Talks are also sometimes given by the staff of the poultry department of one of the local colleges of agriculture and these can prove to be most helpful and interesting. A wide range of facts that have been proved over the years in the

care of birds are often brought out, as is a barrage of useful questions at the end. For the more advanced society a quiz may be held between neighbouring clubs with someone well versed in the art of poultry keeping as chairman. Other talks may be given by various expert fanciers on their particular breeds, the standards laid down, and the art of breeding. Egg shows too are organised and very often the judge goes through all the classes afterwards pointing out the reasons why the first prize eggs were the best plate in a particular class and where the others failed. Finally, at the end of the year, some clubs organise plucking contests and give a prize for the person who can pluck, dress and prepare his bird in the best time and manner.

From these observations, it is clear that there are indeed many advantages to be gained from joining the local poultry society. Knowledge gained here, coupled with knowledge gained through experience helps to make the complete fancier in due course. Experience plays the greatest part in the keeping of poultry and it is only through experience that a good stockman will be produced, able to breed birds to give satisfaction as far as a laying strain is concerned and also as far as exhibition qualities are concerned.

The beginner of today, providing he is prepared to take advice, is the successful fancier of tomorrow. If the basic rules are followed in looking after stock, whatever number, colour or type they may be, an enormous amount of pleasure will be gained from them.

8

THE CHOICE IS YOURS

Before a fancier embarks on keeping poultry he may well have in his mind a rough idea of the breed which he wants eventually to keep. Maybe he has seen them at some local show or maybe it is the breed that relatives or friends have kept successfully for many years. In order to help fanciers, a few comments are given on what are perhaps the most popular breeds, although it must be clearly stated that there are many others which because of lack of space cannot be mentioned.

Before giving a brief description of various breeds, it should be emphasised that many of the breeds have their own particular characteristics and markings. Some birds have five toes whereas others have four, some birds have rose combs whereas others have single combs and indeed some birds are judged entirely on shape whereas others are judged, to a certain extent, on height. If you are contemplating buying show stock which does involve a greater outlay of money than buying normal everyday layers, it is a good idea to take an experienced fancier with you in order to ensure that the birds you are buying possess the right characteristics and, above all, are of the right type.

Ancona
The Ancona is a light breed which is a good layer of white or cream eggs. It is a bird which is beetle green in colour with V tipping at the end of each feather. It is available in either the rose–comb or single–combed variety and is generally a most useful breed to keep.

Hamburgh

As a breeder for many years of this variety it is perfectly obvious that the author must include it. Hamburghs are available in five colours the most popular of which are the silver spangled followed by the gold pencilled. Other colours are blacks, silver pencils and gold spangles. The Hamburgh is a light variety, is not prone to broodiness and lays white eggs. It is such a tremendous layer that in bygone days they used to be called 'everyday layers'.

Indian Game

This is a heavy breed that lays tinted eggs although not many in number. It is a breed that should be seriously considered if the object is to keep birds for the table.

Leghorn

Undoubtedly this must be on the top of any list as far as laying birds are concerned. Leghorns are available in several colours the most popular of which are black or white. They are a single-combed bird with white lobes and undoubtedly one to fill any egg basket with good sized white eggs.

Maran

If the object is to keep birds to lay eggs for the show bench then Marans must be included. They are a heavy breed which means they can also be used for the table but their main function is to lay chocolate-brown eggs. The eggs, although few in number, are usually of a good shape and size. Marans are classed as a barred breed with barring across each feather of a dark/light grey.

Minorca

Another bird that should be on the top of any list should the requirement be white eggs for the show bench. Minorcas are generally found in black only and are certainly a graceful type of bird with single comb and large white ear lobes. Good layers, they are seldom prone to broodiness.

Rhode Island Red

A very popular choice for the fancier who is starting to keep poultry for the very first time and is interested in a breed which not only will be a good layer but also a reasonable table bird. The Rhode Island Red, a good layer of light brown and, in some cases, brown eggs fulfils both these requirements. Although it is prone at times to broodiness it is nevertheless a bird that is good to manage and not flighty.

Sussex

Again a very popular choice particularly the most popular colour of Sussex which is the Light Sussex. A layer of good sized tinted eggs it fulfils the requirements of a commercial laying bird and a useful table bird as well. A good old British breed which is always worthy of a place in any fancier's yards.

Welsummer

Rather an outsider, this bird of Dutch origin and layer of good brown and often deep-brown eggs must be considered by the fancier who requires eggs for the show bench. A bird that is rather striking in colour with its rich golden brown and chestnut plumage.

The breeds listed above are those which tend to be the most popular and good to manage as far as the beginner is concerned. It must be stressed that there are many other breeds of birds that are equally suitable for the fancier to begin his hobby with which are not listed. For instance, such breeds as Australorps, Dorkings, Old English Game, Orpingtons, Plymouth Rocks and Wyandottes could well be considered. In addition, there are the many rare breeds which rely on the support of many fanciers in order that they may be preserved.

There are also the various breeds of bantams. Most bantams, of course, have their counterparts in large fowl. However, there are certainly two pure bantam breeds that should be mentioned, namely Belgian and Sebright bantams. Both these breeds are most attractive, being classed as ornamental birds, and have given many fanciers much pleasure.

In contrast, there is one breed that is often classed as a bantam but which is in fact officially a large fowl and that is the Silkie. This breed, certainly intriguing, is one in which the birds have five toes and blue flesh. Silkies cannot be regarded as a breed to produce eggs, but they can be very highly respected as a breed prone to broodiness, and provide some of the best mothers, especially when crossed.

Breed Clubs

Most of the breeds are covered by active breed clubs and the rarer breeds and those that rely on a handful of fanciers preserving them are catered for by the Rare Breeds Society. Whether from a commercial or exhibition point of view secretaries of all breed clubs are always in a position to help new fanciers of their breed.

Good Luck!

INDEX

DAMAGE NOTED